Creating Organic Standards in U.S. States

Creating Organic Standards in U.S. States

The Diffusion of State Organic Food and Agriculture Legislation, 1976–2010

Samantha L. Mosier

LEXINGTON BOOKS
Lanham • Boulder • New York • London

Published by Lexington Books
An imprint of The Rowman & Littlefield Publishing Group, Inc.
4501 Forbes Boulevard, Suite 200, Lanham, Maryland 20706
www.rowman.com

Unit A, Whitacre Mews, 26-34 Stannary Street, London SE11 4AB

British Library Cataloguing in Publication Information Available

Library of Congress Cataloging-in-Publication Data

Names: Mosier, Samantha L., author.
Title: Creating organic standards in U.S. states : the diffusion of state organic food and agriculture legislation, 1976–2010 / Samantha L. Mosier.
Other titles: Creating organic standards in US states | Creating organic standards in United States states
Description: Lanham : Lexington Books, [2017] | Includes bibliographical references and index.
Identifiers: LCCN 2017011601 (print) | LCCN 2017014073 (ebook) | ISBN 9781498554411 (Electronic) | ISBN 9781498554404 (cloth : alk. paper)
Subjects: LCSH: Natural foods—Standards—United States—States—History. | Natural foods industry—Law and legislation—United States—States—History. | Technological innovations—Law and legislation—United States—History. | Organic farming—Law and legislation—United States—States—History. | Natural foods—State supervision—United States—History.
Classification: LCC KF3864.85 (ebook) | LCC KF3864.85 .M67 2017 (print) | DDC 344.7304/232—dc23
LC record available at https://lccn.loc.gov/2017011601

∞™ The paper used in this publication meets the minimum requirements of American National Standard for Information Sciences—Permanence of Paper for Printed Library Materials, ANSI/NISO Z39.48-1992.

Printed in the United States of America

Contents

List of Tables

List of Figures

Acknowledgments

The completion of this book would not have been possible without the help and support of many. First and foremost, I would like to thank my dissertation committee for the support in initially investigating the subject of state-level organic food and agriculture policy. Susan M. Opp, Dawn Thilmany, Kyle Saunders, and Charles Davis provided valuable feedback throughout the process of working on this book. I would also like to extend gratitude to the many individuals who aided in the data collection process including Dennis Cardoza, Jake Crawford, Jerry Curry, Barry Epstein, Sam Farr, Grace Gershuny, Mrill Ingram, Sarah Johnston, Tony Kleese, Mark Lipson, Miles McEvoy, Stephanie Ritchie, Cameron Smoak, Enid Wonnacott, David Zuckerman, staff at the University of Georgia Cooperative Extension, and the many other state agencies, legislative services, law library, and archive staff members that I came in contact with during the research process. Finally, I would like to acknowledge Cailan, Gibson, Marshall, and Wednezday for their unconditional support.

Abbreviations

AMA	Agricultural Management Assistance
AMS	Agricultural Marketing Service
CAC	county agricultural commissioners
CASFS	Center for Agroecology and Sustainable Food Systems
CCOF	California Certified Organic Farmers
CDF	California Department of Finance
CDFA	California Department of Food and Agriculture
CDHS	California Department of Health Services
COFA	California Organic Farmers Association
COROS	Common Objectives and Requirements of Organic Standards
EHA	event history analysis
FAO	Food and Agriculture Organization
FTC	Federal Trade Commission
GDA	Georgia Department of Agriculture
IFOAM	International Federation of Organic Agriculture Movements
NOFA	Northeast Organic Farming Association
NOP	National Organic Program
NOSB	National Organic Standards Board
OCIA	Organic Crop Improvement Association
OFPA	Organic Foods Production Act of 1990
OFRF	Organic Farming Research Foundation
OTA	Organic Trade Association
USDA	United States Department of Agriculture
VOF	Vermont Organic Farmers
WACO	Western Alliance of Certification Organizations

Chapter One

Approaching the Study of Organic Policy Diffusion

The United States represents the single largest organic market globally.[1] In 2014, U.S. organic sales reached 35.9 billion representing approximately 5-percent of U.S. food sales (FiBL and IFOAM 2016). The demand for organic products in the United States began in the 1960s and has grown steadily since. Organic produce (i.e., fruits and vegetables) and dairy were, and remain, the largest sector of organic products for sale. While organic consumers are often wealthier, well-educated, and Caucasian (Dettmann and Dimitri 2007), demand for organic products is driven by a wide range of salient consumer issue groups including environmentalists, food phobics (i.e., food safety), healthy eaters, welfare enthusiasts, and hedonists (Bonti-Ankomah and Yiridoe 2006; Davies, Titterington, and Cochrane 1995; Hughner et al. 2007). Many of these modern organic consumer groups coincide with social movements, such as the environmental movement, that initially influenced the emergence of the organic market and the demand for regulation in the organic food and agriculture sector.

The present U.S. organic food and agriculture market is regulated by the United States Department of Agriculture (USDA) National Organic Program (NOP), which is housed under the Agricultural Marketing Service division. The NOP sets the standards for the production and labeling of organic goods across the nation. However, the federal government was not the first governmental entity in the United States to pursue the regulation of organic food and agriculture. Starting in the 1970s, state governments were the first to enact organic regulatory policies. Oregon was the first state to pursue organic regulations with the adoption of administrative rules in 1973. The New York State Assembly followed shortly thereafter with the adoption of an assembly bill in 1976, which requested the New York Department of Agriculture to meet regarding the development of organic regulations. By 2010, a total of thirty-eight states adopted organic food and agriculture legislation with an additional four states having organic regulations established solely through administrative rules.

There is a rich history documenting the development of the federal law, initially established under the national Organic Foods Production Act (OFPA) in 1990, but little is known about the role of U.S. states in the development of organic food and agriculture policy and standards. The main theme of this book is to describe how and why organic regulatory legislation was adopted among U.S. states from 1976–2010. As such, this book serves as both a history and exploration of causes and consequences of state's adopting organic regulatory legislation. As a history, this book details when and which states adopted legislative bills regulating the organic food and agriculture market. It also provides information for when states sought to terminate state statutes on organic food and agriculture. Three case studies are featured that provide in-depth analysis for why states adopted organic regulatory legislation and the lessons to be derived from each scenario. To date, no other organic policy study has comprehensively identified the history of policy adoptions among U.S. states. Moreover, there is limited investigation that captures state-federal interactions and struggles to regulate a growing market. The adoption and implementation of the OFPA in 1990 demonstrates the dynamic between the state and federal level for how best to regulate the organic market. This struggle continues as consumers and producers seek to grow the organic market beyond just food.

As an exploration, this book is instrumental for understanding the motivations and frameworks underlying U.S. state organic legislation. The diffusion of organic legislation in the United States spans across four decades and occurs before and after federal preemption in the policy domain. The adoption and reinvention of state organic policies aid in understanding the varying approaches to organic production, the multiple approaches to regulating the market, and identifying future prospects for organic production and sales in an ever-changing policy environment. The current U.S. national standards for organically labeled goods has been determined by a lengthy political process that involves organic consumers, organic farmers, conventional agricultural interests, bureaucrats, and politicians. Organic food and agriculture has been defined through both legislative bills and the administrative rule-making process. The definition of organic food and agriculture varies and can vary dramatically depending on who is being asked. From a legal perspective, there are several different definitions for organic, including specific definitions for organic, organic agriculture, organic agricultural practices, and organic production. In addition, understanding the permissible list of practices and substances requires references to the National List of Allowed and Prohibited Substances, which identifies substances and practices that can and cannot be used in the production or processes of organic crops, livestock, or consumer products. A consumer friendly definition, according to the USDA NOP, is the following:

> Organic is a labeling term that refers to agricultural products produced in accordance with Organic Foods Production Act and the NOP Regulations. The principal guidelines for natural systems and that integrate the parts of the farming system into an ecological whole. Organic agriculture practices cannot ensure that products are completely free of residues; however, methods are used to minimize pollution from air, soil and water. (USDA NOP 2012)

This lengthy and somewhat ambiguous definition outlines the complex meaning of organic to where there is no easy or precise meaning to what items constitute as organic. Indeed, our current conception of organic and the structure of regulation are highly influenced by more consumer-oriented policies developed in Oregon and California in the 1970s. Organic associations, such as the Organic Trade Association, would later refine the conception during the NOP proposed and final rule development around the turn of the twenty-first century (Dimatteo and Gershuny 2007).

Exploration of diffusing state-based organic regulations can point to a number of causes underlying challenges in developing national policy and regulating the continually expanding market. While market research demonstrates significant increases in organic sales, patterns of organic production and consumer interest in other food production systems, such as local and community-based agriculture, demonstrate significant challenges to the future organic market. The organic market is facing increasing pressures from emerging alternative agriculture and food labeling in the marketplace such as local, fair-trade, GMO-free, and natural (Brush and Link 2012; Costanigro et al. 2014; ERS 2009; Onozaka and Thilmany-McFadden 2011; Onozaka, Nurse, and Thilmary-McFadden 2011; Renner 2012). Historically, natural and organic labels have competed in the market with early organic advocates attempting to distinguish between the two (Knorr 1982 and 1984; Price 1985). However, fair-trade, GMO-free, and local-labeled goods are more recent developments. Some would argue these label strategies are in response to the belief that the organic label has lost some of the connotation of representing small-farm, locally produced, ethical or humane, or sustainable (Adams and Salois 2010; Guthman 2004; Popoff 2010; Rigby and Brown 2003; Raynolds 2000; Yue and Tong 2009; Zepeda and Deal 2009). Indeed, consumers may have different motivations for buying organic goods, as opposed to local, non-GMO, or fair trade goods (Guthman 2004; Loureiro and Hine 2001; Onozaka, Nurse, and Thilmany-McFadden 2011; Raynolds 2000).

As the organic market continues to expand and seek a wider consumer base, there are many questions regarding the competition emerging from other alternative or sustainable markets. In part, emerging competition may be reflective of spatial dynamics to the supply and demand of organic foods. The salience of organic food and agriculture surged in the late 1980s as more Americans became interested in organic foods and concerned about conventional agricultural practices. In turn, organic agricultural production increase through the conversion of land to organic production or the emergence of newly formed organic farms. U.S. organic acreage has doubled since the late 1990s and the number of certified operations increased (ERS 2009; NASS 2014). However, the growth of the industry, measured through both acreage and the number of certified operators, was and continues to be relatively limited to areas primarily in the Western, Great Lakes, and New England regions. For example, while there are thousands of certified operations in California and Washington, there has historically been fewer than fifty certified operations in most Southeastern states (NASS 2014). Miles McEvoy, the USDA Agricultural Marketing Service deputy administrator, refers to this phenomenon as "geographic-specific challenges" (McEvoy 2012). He notes potential causes including regional environmental conditions and poor farmer support outreach through state agencies and universities as underlying causes of low certification figures. Academic researchers confirm geographic constraints of the market by tracing the spatial concentration of the organic supply chain, which has increased over time (Hooker and Shanahan 2012). The concentration of production may impact the market for locally grown foods and small-scale farmers. In turn, the market and governmental forces within certain states and regions may predominantly shift away from organic to focus on other alternative food market systems.

This book documents the diffusion of regulatory legislation among U.S. states and identifies key factors leading to the adoption and reinvention of policies. The primary question driving this project is why do some states adopt organic food and agriculture legislation while others do not? In essence, what factors contributed to the initial adoption and spread of the policy via legislation among thirty-eight states since 1976? As secondary considerations, other questions include how historical development in U.S. states has influenced current organic practices and, if historical developments matter, will organic food and agriculture remain competitive in an ever-changing market? The findings of this study reveal that state adoption of organic regulatory legislation is explained by a complex web of influential socio-economic and political factors. States with more amenable political conditions and a strong organic farming community were often the first to pursue government action. However, rising interest in organic farming in the mid- to late-1980s pushed the issue of organic food and agriculture to the forefront of many state policy agendas. The consequences for contemporary organic practices and consumption patterns are inadvertently affected by when and which states adopted organic policies. Delving into the history of organic policy diffusion among U.S. states can shed some light on how U.S. organic policy is designed today.

POLICY DIFFUSION: A FRAMEWORK TO EVALUATE DIFFUSING ORGANIC POLICY

The policy diffusion framework can help to explain the diffusion of organic food and agriculture regulatory policy by theoretically structuring the analysis to focus on key variables leading to legislative action or reinvention and through the identification of avenues for communicating innovative policy ideas. Policy diffusion research specifically centers on explaining how and why innovative policy ideas spread among governmental entities including elected legislatures and administrative organizations. This line of work is grounded in political science but is truly part of a broader academic interest in the diffusion of innovations that spans across many disciplines including anthropology, sociology, education, communication, marketing and business, geography, economics, public administration, political science, and psychology.[2]

Identifying an Innovation

The identification of an innovative policy can be challenging. There are several definitions for defining innovations within the field. Everett Rogers, widely considered as the father of the diffusion of innovations literature, defines innovation as "an idea, practice, or object that is perceived as new by an individual or other unit of adoption" (Rogers 2003, 12). A similar definition is provided by Jack L. Walker, a seminal political science scholar interested in policy diffusion, where an innovation is "a program or policy which is new to the state adopting it, no matter how old the program may be or how many other states may have adopted it" (Walker 1969, 881). Both definitions indicate that innovations are not necessarily inventions, which is the creation of something new. Indeed, the concept of an invention is identified as being different but closely related to innovation. Invention may precede the process of innovation and diffusion (Fagerburg 2005, 5; Jordan and Huitema 2014; Rogers 2003, 43), but it may also exist without the process of invention, as an innovation only requires the perception or use of a policy being new to the adopting unit.

Policy regulating the organic food and agriculture market, ranging from labeling requirements to the production and manufacturing process, can be considered an innovation. What we know today as "organic" is a term born in response to political, economic, and social responses to new agricultural technologies. Before the emergence of conventional and industrial agricultural practices in the nineteenth and twentieth centuries, traditional farming practices could generally be equated as sustainable and organic in nature (Altieri 1990; Baker 2005). The modern day conception of organic farming and production was in response to innovative technologies in agriculture that would come to define conventional practices. As detailed further in chapter 2, the acceptance of organic policies by U.S. federal and state governments was largely spurred by organic farmers and advocates seeking to protect their livelihood and beliefs from being captured and exploited by conventional food and agricultural producers. Federal policies were already in place to support conventional agricultural research, initially through the Morrill and Hatch Act and later through such mechanisms to protect farmers and large biotechnology companies such as Monsanto. The struggle to create a definition of organic and acceptable practice was in the face of adversity and contemporary developments in agricultural research and development. While not all organic farmers agree with the National Organic Program standards, the policies in place at the state, federal, and international level represent a compromise for how to reconceptualize a once common and traditional practice into a modern day system. As such, organic farming is not a new invention, but the policies surrounding the market are an innovation.

Explaining Diffusing Policies—Modeling and Determinants

The study of policy diffusion specifically examines how and why disruptive, permanent, and fundamentally transformative policies spread from one governmental entity to the next. The policy diffusion literature has roots beginning in the mid-twentieth century but attention was not elevated until the seminal work of Jack Walker and Virginia Gray emerged in the *American Political Science Review* in 1969 and 1973 (see Gray 1973a and 1973b; Walker 1969 and 1973). Today, the field is expansive with nearly one thousand articles ranging in foci from understanding primary determinants to mechanisms of diffusion (Graham, Shipan, and Volden 2013; Shipan and Volden 2012). Most policy diffusion studies focus on U.S. states and less so on other governmental entities such as countries, subnational governments such as cities or counties, or administrative agencies.

Early scholarly attempts to model diffusion in the 1970s and 1980s largely fall into two camps: internal determinant and external determinant models. External determinant models, or what Berry and Berry (2007) refer to as diffusion models, examine how non-adopters are influenced by adopters of an innovation through a number of communication channels.[3] Competition,

Table 1.1. External Determinant Models

Model	*Description*
Regional Diffusion	Examines how states within a region influence each other. Evaluations can be neighbor-based or follow a fixed-region approach.
National Interaction Model	Examines how states interact through national communication networks.
Leader-Laggard	Examines how pioneer states are emulated by other states.
Isomorphism	Examines how states (or other adopting units) look toward states with similar characteristics for adoption cues.
Vertical	Accounts for federalism in the diffusion process.

Information adapted from Berry and Berry (2007).

learning, and pressure from other adopting states may lead to emulation among adopting entities. To date, five different external determinant models have been identified in the political science literature by Berry and Berry (see table 1.1 for outline). A debate common to all these models is how policies are communicated through various channels and the consequences of communication styles. In other words, scholars are still investigating whether emulation among states or adopting units is the result of learning or mere mimicking due to external pressure. Of particular concern is the depth of adoption (see Clark 1985; Glick and Hays 1991) and the consequences associated with coercion and mimicking (i.e., non-learning) adoptions (Sharman 2010; Shipan and Volden 2008; Soule 1999; Weissert and Scheller 2008; Weyland 2007).

The two oldest identified external determinant models, the regional diffusion and national interaction model, originate from the Walker-Gray debate. The regional diffusion model, initially proposed by Walker (1969), suggests that potential adopters are influenced by others that are geographically closest to them. Therefore, the probability of adoption increases as other nearby governments adopt an innovation. Walker developed a tree model of diffusion that has two basic steps for tracing the spread of innovation. First, at the base of the tree model there is a set of pioneer states, or the most innovative states, that compete and emulate each other at a national level. For each region in the country, there is at least one pioneer state. According to Walker, "States like New York, Massachusetts, California and Michigan should be seen as regional pace setters" (Walker 1969, 893). Second, there is a set of followers for each region, some considered quick adopters and others considered laggards that usually only adopt a given innovation after the regional pioneer state has had success with the adopted innovation. This emulation occurs because, after observation of the early-adopting state, the perceived risk associated with departing from the status quo policy or program is diminished. These follower states typically have lower innovativeness scores than their regional pioneer state. One assumption of the model is that follower states will not look outside their own region for innovation cues. This assumption has been challenged and is a key focal point for improving upon regional models of diffusion. A number of scholars suggest a decline in regional diffusion because of rapid technology advances that make communication and travel quicker than they were a century ago (Boekmke and Skinner 2012; Shipan and Volden 2012; Weyland 2007). Nevertheless, some evidence suggests that neighbor influence may be stronger or weaker depending on the stage the state is in during the policy process (Cohen-Vogel and Ingle 2007). Therefore, regional diffusion models may present limitations in explaining diffusion patterns among states or governmental entities, but cannot be altogether disregarded.

The national interaction model overcomes some of the limitations of the regional model by suggesting states interact in national communication networks, such as professional organizations, where there is free and equal opportunity to interact with individuals from other states or governments. According to Gray and as stated by Berry and Berry, "the probability a state will adopt is proportional to the number of interactions its officials have had with those that have already adopted" (Gray 1973a; Berry and Berry 2007). Gray's observations for three issue areas (education, welfare, and civil rights) lead her to conclude adopters influence non-adopters on a national level, thereby resulting in an s-shaped curve[4] in cumulative adoptions when plotted (see figure 1.1). Scholars that have used the national interaction model generally find that national platforms for state policy-maker interaction may lead to

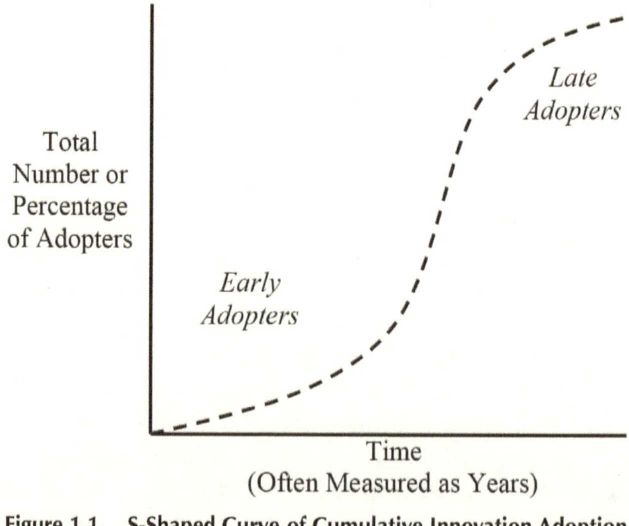

Figure 1.1. S-Shaped Curve of Cumulative Innovation Adoption
Created by the author.

diffusion but a nonrandom pattern of diffusion (Balla 2001; Glick and Hays 1991; Menzel and Feller 1977).

Beyond the regional diffusion and national interaction model, three other external determinant models have been identified—leader-laggard models, isomorphism models, and vertical models. Closely related to the regional diffusion model are the leader-laggard models. Leader-laggard models posit that there are pioneer states that other states seek to emulate. Walker's (1969) tree model of diffusion contains elements of both regional diffusion and leader-laggard models by assuming regional pioneers. However, leader-laggard models are not always spatially confined and can encompass the communication channels present in vertical diffusion, national interaction, and isomorphism models (Berry and Berry 2007).

Studies that use the isomorphism model examine how states or countries with similar characteristics may emulate each other. Similarity characteristics can include population size, ideology, or economic conditions. Regionalism can play a factor in isomorphism models as states or countries in the same regions could share similar social, political, or economic values (Brooks 2005; Nicholson-Crotty 2004; Weyland 2004).

Finally, the vertical influence model accounts for federalism in the diffusion process. In the U.S. context, diffusion researchers have sought to explain how localities (including cities and counties), states, and the federal government may all influence each other. Initially, it was argued state influence on federal adoption was limited (Rose 1973). However, both bottom-up and top-down influences have been observed in the United States in the past several decades (Allen, Pettus, and Haider-Markel 2004; Boeckelman 1992; Daley and Garand 2005; Gray, Gray, and Williams 1981; Karch 2006; Shipan and Volden 2006; Welch and Thompson 1980). From a bottom-up perspective, Shipan and Volden (2006) identified two particular effects local entities could have on state adoption of antismoking policies. A snowball effect occurs when more local adoption leads to momentum for state adoption. A pressure valve effect occurs when local adoption minimizes pressure for state adoption. Interest groups and state legislature capacity determine which effect is observed.

Comparatively, internal determinant models shift the focus from events and conditions occurring outside of adopter-to-adopter characteristics. In essence, internal determinant models focus on the adopter characteristics, including innovativeness and affluence, as the primary explanation for diffusion. Research examining internal determinants of diffusion has reached many of the same conclusions summarized nearly three decades ago by Savage (1985). Wealth, capacity, and size impact the adoption decision (Bailey and Rom 2004; Berry and Berry 1990).[5] In addition, election proximity, problem severity, policy entrepreneurs, and interest groups/advocacy coalitions have been identified as influencing internal conditions necessary for fostering adoption (e.g., see Haider-Markel 2001; Jacobsson and Lauber 2006; Karch 2007; Mintrom 1997). All of these variables have been identified in the broader policy process literature as critical to the passage or modification of policies. For example, Kingdon (2003) indicates the significance of policy entrepreneurs (also see Baumgartner and Jones 2009) and perceived problem severity in policies reaching high agenda status or passage. However, there is mixed evidence for how other characteristics, such as general innovativeness[6] or political culture, influence adoption or adoption proneness (see Savage 1985).

Unified Models of Policy Diffusion

While most early studies separated and evaluated internal and external conditions in isolation, modern advancements in the field have been made under a unified model of diffusion.

Berry and Berry's (1990) unified model of policy diffusion, based on event history analysis, that scholars examined both internal conditions and external conditions under a singular lens of analysis. Berry and Berry claimed a "critical conceptual weakness" existed in the diffusion literature because of the separation of the external[7,8] and internal[9] determinant models. In isolation, each model could not fully explain diffusion and can only offer one potential explanation (Berry and Berry 1990). However, if the two models were unified, scholars would be better equipped to predict adoption influences.[10] Their approach is methodologically grounded in event history analysis (EHA) but is theoretically grounded in Mohr's (1969) work on organizational innovation. EHA is used in biological and other social science research to account for risks and probability of certain event occurrences. Accordingly, EHA is incorporated into the unified model because of its ability to increase the methodological rigor and "substantive relevance of research findings" of policy diffusion research (Berry and Berry 1990). In particular, EHA can reduce spurious relationships by controlling for regional effects, allow for pooled cross-sectional data analysis, and adequately predict adoption probability. Prior diffusion studies did not control external/internal influences and used traditional cross-sectional methods.[11]

Mohr's work on organizational innovation was also incorporated into unified diffusion model because of its ability to drastically improve the diffusion research. Mohr developed and confirmed a three-dimensional hypothesis that predicted "innovation is directly related to the motivation to innovate, inversely related to the strength of obstacles to innovation, and directly related to the availability of resources for overcoming such obstacles" (Mohr 1969). Berry and Berry argued that Mohr's work could integrate elements of regional and internal determinant models of diffusion. The motivation, obstacles, and resources presented to politicians account for internal determinants. Policy adoptions in other states provide information, an external resource, and reduce uncertainty, one perceived obstacle. To test the unified model, Berry and Berry examined state lottery adoptions by developing eleven hypotheses divided into three categories—motivation, obstacles, and resources. What they found confirmed the utility of a unified diffusion model and Mohr's evaluation of innovation in state organizations. Specifically, they found support of regional influence, fiscal health, and election proximity as internal determinants for state lottery adoption. The unified model helped to bridge the significance of both internal and external conditions simultaneously.

The unified model developed by Berry and Berry (1990) not only closed a theoretical gap but a methodology gap, as well. The EHA unified model could account for year-to-year variations in external and internal variables that influence adoption decisions. Only minor adjustments of independent variables within the model have been made, such as region (e.g., see Haider-Markel 2001; Mooney and Lee 1995, 1999, and 2000), addition of vertical variables (e.g., see Allen, Pettus, and Haider-Markel 2004), and the examination of policy change or reinvention (e.g., see Boehmke and Witmer 2004; Clark 1985; Glick and Hays 1991; Hays 1996; Mooney and Lee 1995). The EHA model and Berry and Berry's theoretical approach is the most utilized and cited diffusion model since 1990.[12] It is also the guiding framework for the analysis of organic regulatory policy diffusion.

Mechanisms of Diffusion

In a complementary vein of research, diffusion scholars have also sought to disentangle how policies diffuse. Four mechanisms of diffusion are identified including learning, imitation, competition, and coercion (Shipan and Volden 2008). Learning refers to the process of states observing the actions and experiences of others to determine whether a new policy idea is acceptable for adoption (Berry and Baybeck 2005). It is also cited as a driver for changes in comprehensiveness and content of policies during the diffusion process. Learning insinuates that states will customize a policy according to local needs when searching for a solution to a problem. Imitation refers to the process of adopting a policy without consideration for how the policy performed. Imitation is evident when the language of policies is identical or nearly identical across adopting entities including grammatical mistakes. It is believed that imitation occurs in size and leader-laggard based scenarios, where smaller cities or communities are more likely to adopt a policy if a larger neighbor adopts the same policy (Shipan and Volden 2008).

As a third mechanism, competition is similar to the learning mechanism, in that governments look to the actions of others (Berry and Baybeck 2005; Boehmke and Witmer 2004). However, competition can largely be viewed within the context of race-to-bottom or convergence-at-the-top discussions. The effect of economic externalities, both positive and negative, can influence the decision of governments to pursue a policy. Boehmke and Witmer (2004) find diffusion occurring through economic competition is more likely to lead to expansion compared to policies diffusing through learning. This new direction of mechanism research is significant to understanding the evolution of innovative policy ideas.

Finally, coercion describes a diffusion process whereby governments are pressured, either by peers or a higher authority to pursue a particular policy. Within the U.S. context, vertical pressures are more likely to occur than horizontal coercion. The federal government can use grants or enact preemptive and partially preemptive policies to eliminate the need for state policies or demand state political action in a given policy area.

Of the four mechanisms, learning and economic competition are cited as the two most common explanations for how policies spread among governmental adopters. Scholars have found particular mechanisms occurring at certain times and phases of the diffusion process (see Karch 2007; Kwon, Berry, and Feiock 2009; Shipan and Volden 2012). Later adopters are believed to benefit from cumulative effects of learning from earlier adopters. However, later adopters may also be more likely to face coercive pressures should they resist adopting a policy. Earlier adopters, it is believed, may be more likely to be driven by economic competition. As the literature progresses, this area of policy diffusion research is being more closely linked to understanding changes in policy content over time. In particular, assessment of policy attributes and content are examined. This may explain why learning leads to later adopters having more comprehensive and complex policies initially, compared to earlier adopters.

Policy diffusion provides a theoretical framework for exploring how and why states adopted regulatory policies for organic food and agriculture. The framework presents the opportunity to historically construct the motivating factors beyond initially adopting organic policies and desires to reinvent the content over time. In addition, policy diffusion can uncover how the policy was communicated among states and determine how the policy diffused from the subnational to the national level. Tracing the process of diffusion leads to insights for how modern day policy is constructed and the challenges that are currently found in the organic market.

PLAN OF THE BOOK

The remainder of the book covers the history of organic agricultural production and highlights why and how organic regulatory policies diffused among U.S. states. Chapter 2 covers the history of organic farming in the United States with particular attention to policy developments and the contemporary structure of the National Organic Program. The chapter begins by revisiting why organic policies are innovative in nature. The practice of organic farming is not a new invention or practice. Historically, farmers practiced organic farming for millennias prior to the modern era of farming and food production. What altered the perception of organic and led to government developed organic policies were changes in agriculture sector. Next, the chapter covers a brief history of organic agriculture in the United States. This history covers social movements surrounding organic farming, details the dates in which states adopted organic policy, and covered the development and current structure of the National Organic Program. The chapter concludes with general expectations for how organic policy diffused in the United States.

In chapter 3, the specific causes and determinants for the diffusion or adoption of organic food and agriculture legislation by U.S. states are evaluated. The chapter provides a broad framework for how organic policy diffused and understanding how particular socio-economic and political conditions explain state adoption decisions. It is the results from this broad perspective that guide the state case evaluations of California, Vermont, and Georgia in chapters 4, 5, and 6. Each case study evaluates a different time period for initial adoption of organic regulatory policy and highlights the reinvention of organic policy over time. California, a leader in adoption, demonstrates the power of political and economic conditions that drove demand for government intervention in the organic sector. In addition, the case of California points to the power of state advocates, nested with a broader national organic network, to push for federal policy and later challenge the direction of the National Organic Program. Vermont, whose state legislature initially pursued organic regulations in 1990, provides a counterpoint to the organic policy emerging on the West Coast. Constructed from a producer perspective, the Vermont regulatory law was structured to challenge the consumer perspective law developed in California and transplanted to the federal level. Finally, the case of Georgia demonstrates why some states were late adopters. State lawmakers, at the request of the Georgia Department of Agriculture, adopted organic regulatory policy ten years after the adoption of the OFPA. In part, the late adoption in Georgia is explained by delayed federal action. However, Georgia's actions are also reflective of geographic factors. States located in the Southeast possess a political and social culture that was not amenable to such regulations.

Finally, chapter 7 summarizes the findings from this study and contemplates the future of organic food policy. The results demonstrate that the adoption of a new and innovative policy, such as organic food and agriculture regulation, is dependent on the conditions of a specific time and context. While this finding is not extraordinary, the context of adoptions and process of reinvention does present clear evidence for how national organic standards have developed. Moreover, the historical evidence does suggest a number of potential reasons and solutions for how organic will function in an ever-changing food and agriculture market.

NOTES

1. European and North American consumers represent about 90 percent of all global demand (FiBL and IFOAM 2013).

2. Everett Rogers's *Diffusion of Innovations* (2003) provides a comprehensive overview of the diffusion literature in the social sciences. It should be noted that Rogers classifies political science and public administration as minor disciplines in the diffusion literature.

3. The dependent variable in external determinant models is most often binary (e.g., adoption, non-adoption).

4. The s-shaped curve demonstrates how the cumulative total number of adopters increases over time as the interaction between adopters and non-adopters occur. It is considered as the normal distribution of the cumulative number of adopters when plotted. It has been observed in all disciplines that study diffusion. Gabriel Tarde (1903), a forefather of sociology and social psychology, was the first to observer the s-shape curve of adoption over time.

5. Fairbanks's (1979) morality study, Regens's (1980) energy policy study, and Rosebaum's (1976) land use study are noted exceptions for finding a positive relationship between affluence and adoption likelihood.

6. It is significant to note that only one article has been published since 1985 on innovativeness as a state characteristic. Boehmke and Skinner (2012) reexamine Walker's (1969) original scores and discover drastic changes in both aggregate and individual levels of innovativeness in the states.

7. Berry and Berry (1990) outlined three methodology approaches to regional diffusion explanations including factor analysis (see Walker 1969; Canon and Baum 1981), assessing the relationship between adopters and non-adopters (see Lutz 1986),

and surveying state officials on the influences of adoption (see Freeman 1985; Grupp and Richard 1975, Light 1978; Menzel and Feller 1977).

8. At the time, Berry and Berry did not include vertical diffusion studies, those studies focusing on local-state-federal influence (e.g., see Boeckelman 1992; Daley and Garand 2005; Karch 2006; Shipan and Volden 2006), as part of regional diffusion but would later include vertical diffusion studies as part of a broader "diffusion models" category (Berry and Berry 2007).

9. At the time of publication in 1990, internal determinant models used cross-sectional analysis with the dependent variable either being year adopted or whether a state adopted a policy by a certain date.

10. Berry and Berry (1990) indicated there were a number of studies that evaluated both internal determinants and regional diffusion models. However, scholars never combined the two models and kept evaluation of each model separated.

11. Prior to Berry and Berry's (1990) unified model, no year-to-year variation of independent variable data was utilized in the diffusion literature. One potential explanation for avoiding pooled cross-sectional data is the limited variance in the dependent variable (e.g., adoption/non-adoption) on a yearly basis. However, Berry and Berry do not find low adoption rates, as adoptions are considered unusual and infrequent events, problematic for their analysis on state lottery adoption.

12. Boushey's (2010) epidemiology model is one unified approach that challenges some of the underlying assumptions of incremental evaluation and diffusion. His work looks at rapid and sudden outbreaks of diffusion that do not conform to the normal s-curve distribution.

Chapter Two

A Revolutionary Idea

Organic as an Innovation

Organic agriculture is a not an innovation, but policy regulating the market is an innovation. Constructing the meaning and parameters for what "organic" means is something that could not be achieved without the emergence of industrial and conventional agricultural practices to provide comparison. This chapter provides an overview of the history of organic food and agriculture in the United States. Of the thirty-five-year period analyzed in this study, it becomes clear that the diffusion process operated in distinctive periods marked by changing perceptions of organic agriculture. The diffusion of organic regulatory legislation across thirty-eight states occurs amid a backdrop of rising concern over conventional agricultural practices, delayed federal action, emerging international conflict on agricultural norms, and a changing USDA constituency base.

This chapter unpacks broader historical trends of organic food and agriculture policy in the United States. The chapter begins by briefly revisiting the concept of organic as innovation. Next the history of organic policy action is assessed through a pre-OFPA era and post-OFPA era. The national government was absent in regulating the market in the 1970s and 1980s. While there were attempts under the Carter administration to support and study the burgeoning market, the election of Reagan ended any concern for organic agriculture. However, rising consumer concern about the dangers of conventional agriculture and increasing adoption of state legislation pushed federal lawmakers into action that would result in the OFPA and a new era of organic policy in the United States. Finally, the chapter concludes with how the policy diffusion theoretical framework is applied to the area of organic regulatory policy. General expectations are identified.

ORGANIC AS AN INNOVATION

As briefly discussed in chapter 1, organic agriculture is not an innovation, but organic food policy is an innovation. What we know today as "organic" is a term born in response to political, economic, and social responses to new agricultural technologies. Before the emergence of conventional and industrial agricultural practices in the nineteenth and twentieth centuries, traditional farming practices could generally be equated to sustainable and organic in nature (Altieri and Carroll 1990; Baker 2005). The modern day conception of organic farming and production was in response to innovative technologies in agriculture that would come to define conventional practices.

Industrial and conventional are two terms that are used interchangeably to describe modern agricultural practices. While some may argue that the two terms are different, the general premise of both concepts rest on highly mechanized and technology-driven practices. In the nineteenth century, advancements in technology largely improved the machinery available to farmers. This development was particularly important for mid-western farmers, who specialize in grain and livestock production (Schwieder 1990). The Morrill Act of 1862 and Hatch Act of 1887 aided in the new technological advancements to help farmers increase productivity and yields. Advancements in technology during the twentieth century are truly where our modern conception of conventional agriculture was born. The post–World War II environment in agriculture made it acceptable to use chemical and synthetic pesticides and fertilizers. Later, biotechnology developments, leading to patented seeds, are also a hallmark of conventional practices in addition to such practices as monocropping and fossil fuel intensive labor practices. To some, there are dire consequences resulting from conventional agriculture including the decline of the family farm, rising socio-economic inequalities, soil and environmental degradation, and risks to human health (Norberg-Hodge, Goerin, and Page 1993; Lobao and Meyer 2001). The general perception of conventional agriculture is one of disconnect from the natural environment.

Organic farming, in turn, is largely considered connected to the natural environment and a direct rejection of industrialized and conventional agriculture. As Niggli indicates, "Organic farming was initially characterized, in part, by a sometimes fierce rejection of the so-called conventional farming techniques characteristic of mainstream agriculture" (Niggli 2007, 73). Early proponents of organic practices, such as Eve Balfour, Albert Howard, and Rudolph Steiner, emphasized holistic farm management and production instead of an emphasis on narrowly defined research (Stinner 2007, 40–42). Values attached to organic farming include health, ecology, fairness, and caring (Sligh and Cierpka 2007). Each component instills images of sustainability, naturalness, local, and environmentally friendly. Often absent in the common perception of organic agriculture is the use and utility of scientific research. Organic agriculture has gained just as much from science as conventional agriculture through such field findings on weed ecology or crop quality (Stinner 2007). However, the use of technology may be more in line with ecological harmony compared to conventional practices.

The perceptional differences between organic and conventional agriculture may be stark but each is representative of two-sides of the same coin. Both emerged in the response to concerns about feeding an increasing population and addressing poor soil quality. Industrial agriculture took the approach of beating nature into submission while organic agriculture sought to work within the bounds of nature, which is truly just an extension of traditional farming (Sligh and Cierpka 2007, 30). The industrial approach gained mainstream acceptance earlier in the twentieth century while organic did not. Organic agriculture has been labeled as kooky, antiquated, non-innovative, and detached from real world demands. Indeed, Rogers (1957 and 2003, 193–94) once classified a farmer as an innovative laggard because the farmer refused to apply newly developed agricultural chemicals, such as 4-D weed spray, on his crops. However, the laggard farmer was not really an innovative laggard. By the 1980s, organic farming and consumer interest in organic foods soared. In hindsight, Rogers (2003, 194) classified the farmer as a "superinnovator."

The policies that emerged to support and define conventional food and agriculture helped shape the research and perception of organic food and agriculture. For organic, defining acceptable practices and obtaining government support for research and development was much more difficult but would not be possible without government involvement and support for conventional agricultural practices first. Organic farming and food is not a new practice or idea, but the modern conception and subsequent governmental policies are radically innovative.

A BRIEF HISTORY OF ORGANIC FOOD AND AGRICULTURE POLICY IN THE UNITED STATES

The history of organic food and agriculture policy in the United States is a rich and complex story. From 1973–2010, forty-two states adopted some form of law regulating or permitting the development of regulations for organic food and agriculture (see tables 2.1 and 2.2). Many of these states adopted organic regulations prior to the congressional adoption of the OFPA in 1990. Yet, twenty-one states adopted organic law on or after 1990. The delayed implementation of the National Organic Program, established under the OFPA, until

Table 2.1. States Adopting Organic Legislation (1976–2010)

State	*Year of Initial Adoption*	*Year(s) of Amending Adoption*
New York	1976	
California	1979	1982, 1990, 1991, 1992, 1993, 1994, 1995, 1996, 1997, 1998, 1999, 2002, 2003, 2005, 2009, 2010
Connecticut	1979	1980, 1981, 1998, 2002
Maine	1979	1985, 1989, 1991, 2003
Minnesota	1985	1988, 1990, 1996, 1997, 1999, 2003, 2005, 2009
Montana	1985	1999
New Hampshire	1985	1986, 1991, 2002, 2003
Washington	1985	1987, 1989, 1992, 1995, 1997, 2002, 2010
Nebraska	1986	
North Dakota	1987	1991, 2003, 2009
Iowa	1988	1989
South Dakota	1988	2003
Wisconsin	1988	
Alaska	1989	1998, 2003
Colorado	1989	1993, 2002, 2005, 2009, 2010
Louisiana	1989	
Oklahoma	1989	2000, 2003
Oregon*	1989	1991, 2001
Texas*	1989	1991, 1993, 1995, 1997, 2001, 2003, 2007
Florida	1990	1994, 1995, 2001
Idaho	1990	1991, 1994, 1999, 2000, 2001, 2002, 2007
Kentucky	1990	1994, 1996, 2000, 2002, 2008
Maryland	1990	1992, 2002, 2009
New Mexico	1990	1993, 2001, 2004
Vermont	1990	2003
Virginia	1990	2001, 2003
Arizona	1992	1994, 1996, 1997, 2000, 2002
Indiana	1993	2006, 2008
North Carolina	1993	
Pennsylvania	1994	2010
Nevada	1997	2003
Georgia	2000	2002, 2010
Michigan	2000	2006
Mississippi	2000	2003, 2004
Missouri	2000	2002
New Jersey	2003	
Illinois	2007	
Wyoming	2009	

**States known to have adopted administrative rules prior to adoption of legislation*

Source: Data compiled by author from state legislative databases and law libraries.

Table 2.2. States Adopting Administrative Rules Prior or in Lieu of Legislation (1976–2010)

State	*Administrative Rule Adoption Year*	*Year of Initial Legislation*
Oregon	1973	1989
Massachusetts	1978	
Texas	1988	1989
Ohio	1990	
Rhode Island	1990	
Utah	2007	

Source: Data compiled by author from state legislative databases and law libraries.

2002, led to skepticism and concern among the states for when and if federal organic standards would be established. Of the twenty-one states adopting organic law from 1990–2010, seventeen states adopted organic law during the federal period of delayed implementation from 1990–2002.

The approach to regulating the organic market varies substantially from state-to-state with most states opting to adopt legislation first followed by a set of administrative rules later. Two key components to understanding the organic policy development and variation are distinguishing between adopted legislation versus administrative rules. Legislation and administrative rules may end with a law and policy being established, but each form of law originates from a separate nexus of decision making with varying rules and actors in the decision-making process. Similarly, there is variation in approaches to regulating organic food and agriculture market activities. Regulation may not be strict and command-and-control in nature. Instead, regulatory policy designs may be collaborative by inviting participation from key stakeholders to help identify normative practices and behaviors. The National Organic Program currently operates under a third-party certification and labeling scheme, which is not a traditional or strict command-and-control regulatory approach. Instead, the third-party certification and labeling process is more collaborative and co-regulatory in nature, permitting public-private engagement for implementing the law.

In what will follow, a brief history of organic food and agriculture policy is organized into two development eras for organic food and agriculture policy in the United States. The first era examines development of organic food and agriculture policy from 1973–1989. This era was highly influenced by establishment of the organic movement and private third-party certification programs. The second era focuses on the development of organic policy in a post–national Organic Foods Production Act era from 1990 to present day. It is in the second era that states interacted and engaged with national policy and emerging challenges to organic from other alternative food systems such as local and community-based agriculture.

The Organic Movement and Early Regulation in the States (1973–OFPA 1990)

Early efforts to regulate the organic food began in U.S. states. Attention to the issue of organic farming originates from several movements seeking to improve environmental and human health. While states were the first and only political actor to adopt regulations, the federal government did consider regulatory action for organic products. However, this effort was stunted with changes in political tides of the 1980s. Early efforts can be summarized as an attempt to provide some regulatory guidance as to what "organic" meant, how to appropriately identify organic food, and the development toward comprehensive regulations and standards.

Social Movements for Organic Food

The organic farming movement in the United States has an extensive history dating back to the beginning of the twentieth century. Given the breadth of what organic production means, there are several social movements that are connected to the support and development of the organic movement.[1] According to Guthman (2004, 4), who wrote extensively about the organic social movement in California, there are four broadly conceived but interrelated social movements that influenced the emergence of the organic market in the United States. Of the four social movements Guthman identifies, the environmental movement is cited as having a less direct but complementary influence on the emergence of organic farming and food (Guthamn 2004, 7–9). Environmental concerns directly pertaining to organic farming and food would only emerge in the 1970s and 1980s (Vogt 2007).

Three of the social movements, soil conservation, pure food, and the counterculture movements, had a more direct impact on organic production. Soil conservationists were largely concerned with improving agricultural practices to address the negative consequences, such as poor soil quality, which could result from industrial or conventional agriculture methods.[2] Depressed agricultural prices in the 1930s and poor soil management practices that contributed to the Dust Bowl led to concerns about the direction and consequences of conventional agricultural practices (Guthman 2004, 4–5). For the pure food movement, the range of scholars and journalists within this movement are primarily concerned with the contamination and adulteration of foods. This movement has expanded to include concern over working conditions in food production and manufacturing that were highlighted in such works as Upton Sinclair's *The Jungle* (1906) and Eric Schlosser's *Fast Food Nation* (2001). Finally, the 1960s counterculture influenced organic agriculture through the establishment of communes, food cooperatives, and utopian experiments inspired by back-to-the-land movements (Guthman 2004, 6–7). References to organic agriculture as a fringe, "hippie" practice are the result of the counterculture movement's involvement in producing organic foods. This particular image is often what led to negative perceptions of organic agriculture as a viable industry. Yet, the counterculture movement is responsible for attempts to distinguish the differences between organic and natural foods, a concern that is now reemerging.

While it could be argued that these social movements could be expanded or consolidated, it is evident in each movement that there is a concern for how food is produced, processed, and distributed, as well as its interface with natural resources and human health. It is important to remember that organic farming largely remained at the fringes of the agricultural community and was unpopular in political circles for most of the twentieth century (Blobaum 2010; Ingram and Ingram 2005; Lipson 1998; Youngberg, Schaller, and Merrigan 1993). It was only the late 1980s that the organic market growth "took off" and organic production became mainstream (Rawson 1998). Consumer concerns over chemical residues on foods and an increase in the number of organic farms led to explosive growth in sales starting in 1989. According to an industry survey, retail organic foods sales in 1990 reached $1 billion in the United States (Organic Trade Association [OTA] 2011). Sales would continue to increase by approximately 20 percent each year after the passage of the OFPA. Prior to the 1990s, a patchwork of state and third-party standards regulated a growing market where federal policy was absent.

Early Federal Policy Action

In the 1970s, organic agriculture was not promoted as a viable industry at the federal level. An iron triangle[3] existed between commodity agriculture interests, congressional agricultural committees, and the USDA. The iron triangle maintained support for conventional agricultural practices and resulted in limited federal support for alternative production methods including organic farming (Ingram and Ingram 2005). Earl Butz, the secretary of agriculture from 1971–1976,[4] served as a gatekeeper for the USDA by preventing any action on organic production during his tenure. In an infamous response to proponents of organic agriculture, Butz remarked, "Before we go back to organic agriculture, somebody is going to have to decide what 50 million people we are going to let starve" (Butz 1971). Butz's attitude toward organic food and agriculture was not uncommon and represents why two efforts to regulate and support organic agricultural practices failed at the national level.

One early federal regulatory effort in the alternative market was to regulate the term natural on foods. In 1979, the Federal Trade Commission (FTC) did propose a rule for regulating food advertising, including defining organic and natural foods, but the rule failed to be implemented due to backlash in other nutrition labeling concerns and lack of interest by Congress to regulate the term (Federal Trade Commission 1979, 1980; Obach 2015, 51–52; USDA Study Team 1980, 6). Instead, the FTC would evaluate individual cases to assess deceptive practices. One state, Connecticut, did succeed where the FTC failed. Connecticut's 1979 Public Act No. 379 defined natural and organic foods and banned false labeling or advertisement.

Prior to the 1979 proposed Federal Trade Commission rule, the USDA and Congress demonstrated some support for organic agricultural practices. Amidst pressures from the oil crisis and environmental concerns, Congress commissioned the USDA to do a study on uses of organic waste to improve soil productivity in 1977 (Rawson 1998). In 1980, under the leadership of agricultural secretary Bob Bergland, the USDA published its own commissioned report titled *Report and Recommendations on Organic Farming* and created the Office of Organic Resources Coordinator to oversee organic research and education programs (Rawson 1998; Rich 2008; USDA Study Team 1980). However, the organic resource coordinator position and office would be abolished under the Reagan administration in 1981, effectively ending organic agriculture programs and research in the USDA until the 1990s.

State Legislation and Third-Party Regulations

Despite being kept at the fringes at the national level, organic farmers had success in prmoting organic agriculture policy at the state level. In 1976, New York was the first state to pass organic legislation by requesting the New York Department of Agriculture to consider adopting organic rules. California, Connecticut, and Maine would follow New York's lead in developing organic labeling rules in 1979. By 1990, twenty-six states had passed organic legislation ranging from labeling, certification programs, or agreements with third-party certifiers. However, legislation was not the only approach to creating organic regulations. States like Oregon, Massachusetts, and Texas had active state departments of agriculture that took a broad interpretation of delegated policy-making ability to create organic standards and certification programs. Specifically, the Oregon Department of Agriculture interpreted delegation of authority under food additive and labeling statutes to include the development of organic definitions and standards. This landmark regulatory effort is largely cited as the beginning of organic law in the United States, but, in truth, the regulatory effort was not the result of elected officials seeking to establish a new policy. Instead, efforts at the bureaucratic level is what initially began the discussion of defining what organic meant, how organic food should be identified on food, and what farming and production practices were constituted as organic. The Oregon legislature would not revise the 1973 administrative rules until 1989.

A similar story is told in Texas and Massachusetts. In Massachusetts, the Massachusetts Department of Public Health enacted regulation 105 CMR 520, regulating organic as part of a larger public health effort in 1978. In Texas, the Texas Department of Agriculture established an organic certification program through administrative rules in 1988 without specific delegated authority from the Texas state legislature. A year later, in 1989, under Texas Senate Bill 489, the legislature restricted the amount of fees to be collected for promotion and cost of operating a certification program. The Massachusetts state legislature never addressed the public health developed regulation.

Coinciding with state action in the 1970s and 1980s, several organic farmers' associations were also moving to self-regulate the emerging market. The Rodale Institute, Oregon Tilth,[5] and California Certified Organic Farmers (CCOF)[6] were three prominent organic farming organizations that sought to improve organic practices and influence policy at the state and federal level. In addition, each organization offered organic certification of products.

The earliest certification program was operated and paid for by Rodale Press's *Organic Gardening and Farming* magazine in the early 1970s (CCOF 1988). Founded in 1947 as the Soil and Health Foundation by Jerome Rodale, the institute provided authoritative information on organic management practices to farmers (Rodale Institute, n.d.). The institute expanded in the 1970s with the acquisition of a 333-acre farm in Pennsylvania and dedication to researching sustainable organic methods in addition to providing certification.

The West Coast editor of *Organic Gardening and Farming* magazine, Floyd Allen, oversaw the examination of thirty-four California farms for organic certification. California Certified Organic Farmers' organic certification program would emerge a few years later in 1973 at the direction of Floyd Allen. It would be several years later before Oregon Tilth's first certification program was organized in 1982 by the Willamette Valley chapter (Oregon Tilth N.A.). The requirements for these certification programs did vary, but the standards were similar enough to provide guidance for what was more or less organic in nature. The early efforts by organic farmers to self-regulate had profound implications for the federal regulatory structure that would emerge in the 1990s by pushing for collaboration among farmers and policy makers in developing unified national standards.

The National Organic Foods Production Act and Current Developments (1990–Present)

The national Organic Foods Production Act was passed by Congress as Title XXI of the 1990 Farm Bill. The 1990 bill was not the first federal legislation seeking to regulate the organic market. Three congressional bills were introduced in 1989 but each ultimately failed (Lathrop 1991). State agency associations and several industry groups petitioned Congress in the late 1980s to eliminate problems in the market. These groups were collectively seeking to eliminate the differences in standards and to address newly emerging concerns in the market (1990 U.S.C.C.A.N. 4656, 493–44; Johnson 2008). Newly emerging concerns included a burgeoning international organic market, multi-ingredient organic products, and addressing organic production methods for meat, poultry and seafood. OFPA aimed to eliminate consumer confusion, improve interstate commerce, and solve problems created by differing state and private entity standards by establishing national standards for governing the organic market (7 U.S.C. 6501).

The bill authorized the creation of the National Organic Program and the National Organic Standards Board (NOSB) to be implemented by the USDA Agricultural Marketing Service. The NOP is responsible for setting the standards for organic production, handling, and processing. In addition, the NOP oversees organic certification to ensure compliance with set standards. Private certifiers and state certification programs would have to be accredited under the new national standards to certify organic producers, processors, and handlers. As another key element of the legislation, the NOSB is an advisory committee for setting the standards by which the NOP operates. There are fifteen members of the board that advise the secretary of agriculture, representing farmers and growers, handlers and processors, retailers, environmentalists, resource conservationists, certifiers, and consumer and public interests.

The first NOSB was appointed in 1992 and submitted rule recommendations from June 1994 to September 1996 to the NOP staff (Rawson 1998). The postponed appointment of the board for two years marked the beginning of delayed implementation. Specifically, the years between passage of legislation in 1990 and appointment of the board have been referenced as "the lost years," where there was significant uncertainty on whether the policy would be implemented. The original implementation date was expected to be October 1, 1993, but lack of funds contributed to the delay of implementation (Gershuny 2002; Rawson 1998). However, even after the board was appointed, there still remained questions regarding the ability of the NOP to ever successfully be implemented (Gershuny 2016). After two years of submitted rules to the NOP, the first proposed rule for the NOP appeared in the *Federal Register* on December 16, 1997 (62 FR 65850). The 1997 proposed rule was highly controversial and received a record number of comments.

By the end of the extended public comment period in April 1998, the USDA received over 275,000 comments on the proposed rule. According to Rawson (1998), the Organic Trade Association identified nine specific problem areas with the proposed rules, inferring that a majority of the comments

Table 2.3. Identified Problem Areas with 1997 NOP Proposed Rule

1	The proposed rule would eliminate key concepts (ex: ecosystem health, biodiversity) from the definition of organic agriculture.
2	The proposed rule undermines the authority of the NOSB. The USDA's recommendations on organic practices and materials substituted the recommendations of the NOSB.
3	The proposed rule allows genetically engineered crops and organisms in organic agriculture.
4	The proposed rule permits irradiation of organically produced foods.
5	The proposed rule allows potentially toxic municipal sewage sludge to be used as fertilizer.
6	The proposed rule is too lax in organic livestock production standards.
7	The proposed rules create unnecessary loopholes for the use of synthetic materials to be used in organic production.
8	The proposed rule undermines the integrity of the organic label by giving the USDA the sole authority to decertify organic operations, which could lead to long delays.
9	The proposed rule does not consider land history as part of its certification requirements. Some soils may be used in organic production that is unacceptably contaminated.

Source: Rawson 1998

related to these concerns (see table 2.3). Of particular concern was permitting the Big Three (i.e., irradiation, GMOs, and sewage sludge) into organic production. The concern over the Big Three was raised because the USDA did not explicitly condemn the use of these practices, which proponents claimed create negative environmental and health effects (Manoochehri 1998). Organic farmers and producers saw this as an attempt to destroy the meaning of organic and to corrupt the practice to where large agribusinesses and corporations could enter the market easily.

There are two explanations for the controversy surrounding the 1997 proposed rule including a changing USDA constituency base and international trade dilemmas (related to harmonization of standards) between the United States and Europe. Since the 1950s, the USDA's key constituents were farmers that "eagerly adopted the stream of new technologies that flowed from the research labs of the land grant colleges of agriculture and from private agribusiness firms" (Rawson 1998). However, by the 1990s, the USDA became responsible for food safety regulation along with the Food and Drug Administration and the Environmental Protection Agency. A new domestic constituency emerged that approached conventional agricultural practices with greater caution.

This cautious approach to conventional methods was also present in the international community. The global market has been regulated by a combination of voluntary and compulsory international, state, and local policies and agreements. The International Federation of Organic Agriculture Movements (IFOAM)[7] and the Codex Alimentarius Commission[8] each established voluntary guidelines and standards. Both sets of guidelines are considered international voluntary standards meant to guide states, producers, and consumers to identifying acceptable organic management principles.[9] Compulsory measures are currently in place in eighty-six countries including the United States (FibL and IFOAM 2013). The NOP is considered to be the premier labeling and accreditation system globally and has arrangements with several nations to facilitate trade (NOP 2013a). Yet, there remained much caution in the 1980s and 1990s regarding genetically engineered crops. In the 1990s, the United States was actively working to ensure U.S. agricultural products would be accepted into international markets. The European Union was denying the import of genetically engineered soybeans and corn from the United States on the basis of environmental and health concerns. The 1997 proposed rules were reflective of these precarious policy scenarios. The USDA was deliberative in distinguishing GMOs and synthetic pesticides as being different but equally safe to organic practices.

On May 8, 1998, the USDA announced it would revise the 1997 proposed standards and specified it would not permit irradiation, GMOs, or use of sewage sludge in organic production. Two years later, the USDA published a revised proposed rule on March 13, 2000 (65 FR 13512), and published a final rule on December 21, 2000 (65 FR 80548), in the *Federal Register.* On October 12, 2002, twelve years after the passage of OFPA and nearly ten years after the expected implementation date of 1993, the NOP was officially and fully implemented. Under the final rule, organic production is defined as "a production system that is managed in accordance with the Act and regulations in this part to respond to site-specific conditions by integrating cultural, biological, and mechanical practices that foster cycling of resources, promote ecological balance, and conserve biodiversity" (7 C.F.R. pt. 205.2). This definition reflected the philosophical, ecological, and biological concerns identified by the USDA's new organic constituents and the international community.[10] Since the final rule, the NOP has published nearly nine hundred notifications, rules, and proposed rules to continuously improve and maintain the integrity of the organic program.

A Redefined Role for the States

When the NOP was officially implemented in 2002, the role of the states in organic policy changed dramatically. Some states, like California, were actively engaged in NOP rule making and sought to influence the direction and structure of the national program. California and Alaska even passed legislation in 1998 urging the USDA to reconsider aspects of the proposed rule. Yet, despite a state's decision to influence the development of the NOP, only a few policy options would be available after 2002. According to Miles McEvoy, current deputy administrator of the NOP, there are three options for states to support the national program (Personal Communications, Telephone, 1/29/2013). First, states may apply to operate their own organic program. State organic programs are partial preemption arrangements where states are given the opportunity "to oversee its production and handling operations per the USDA organic regulations," but can "also add more restrictive requirements" if desired (NOP 2011). However, additional requirements must comply and not conflict with the minimum standards. California is the only state approved to operate a state organic program. Thus, certifiers and operations within California must meet the state's requirements and standards to be labeled as organic.

A second role for the states is to become certifying agents. Like private entities, states may seek to certify organic operations. State departments of agriculture are often the government entity that operates as a certifying agent. Fifteen state departments or agencies are currently accredited certifying agents and vary in the certification scopes (see table 2.4). Like other certifiers, state certifiers may specialize in scope by certifying crops, livestock, wild crops, or handling.

Table 2.4. State Departments and Agencies as Certifying Agents as of December 1, 2016

State Agency	*Certification Scopes*
Colorado Department of Agriculture	Crop, Livestock, Wild Crop, Handling
Iowa Department of Agriculture and Land Stewardship	Crop, Livestock, Wild crop, Handling
Idaho State Department of Agriculture	Crop, Livestock, Wild Crop, Handling
Kentucky Department of Agriculture	Crop, Livestock, Wild crop, Handling
Maryland Department of Agriculture	Crop, Livestock, Wild crop, Handling
Montana Department of Agriculture	Crop, Livestock, Wild crop, Handling
New Hampshire Department of Agriculture, Markets & Food	Crop, Livestock, Wild crop, Handling
New Jersey Department of Agriculture	Crop, Livestock, Wild crop, Handling
New Mexico Department of Agriculture	Crop, Livestock, Wild crop, Handling
Oklahoma Department of Agriculture, Food and Forestry	Crop, Livestock, Wild crop, Handling
Oregon Department of Agriculture	Crop, Handling
Rhode Island Department of Environmental Management	Crop, Livestock
Texas Department of Agriculture	Crop, Livestock, Handling
Utah Department of Agriculture	Crop, Livestock, Wild crop, Handling
Washington State Department of Agriculture	Crop, Livestock, Wild crop, Handling

Source: National Organic Program (2016)

The third role for states in supporting the national program is participation in one of two certification cost-share programs. The 2002, 2008, and the 2014 Farm Bills authorized funds for the National Organic Certification (NOC) cost-share program. The Agricultural Management Assistance (AMA) cost-share program received mandatory funding from section 2801 of the Farm Bill's Conservation title as created by section 1524 of the Federal Crop Insurance Act (Johnson 2008). While funding for the NOC ended in FY 2012, both programs offered financial assistance to operators seeking organic certification to offset costs. The NOC offered assistance to producers and handlers in all fifty states where agreed upon, whereas the AMA offers assistance to producers in sixteen states.[11]

ORGANIC DIFFUSION

The organic market has changed dramatically since organic policy first passed in Oregon in 1973 via administrative rule making and in New York in 1976 through state legislative adoption. The policy diffusion framework helps to answer key questions of this study and uncover the dynamics beyond how and why organic legislation spread from state to state. First and foremost, what factors contributed to the initial adoption and spread of the regulatory legislation since 1976? As a set of secondary questions, why were some states early adopters of the innovation? Conversely, why were some states late adopters and pursued regulatory legislation after the adoption of OFPA? And finally, what are the long-term impacts of states being the first to adopt organic regulations in the United States? The diffusion literature focuses on uncovering the dynamics for how and why innovative policies emerge. Moreover, if used appropriately, the approach permits a historical construction that identifies both positive and negative consequences associated with organic regulatory policy first being developed in the states.

Based on an overview of the diffusion literature, there are several conditions of interest to examine the diffusion of organic regulatory legislation. The structure of the analysis is based on previous research within the field of policy diffusion and agricultural studies. Namely, the analysis is structured on a host of internal and external determinants likely to influence states to adopt organic legislation. A total of twelve determinants are evaluated within the study using a mixed-methods approach for analysis with the statistical results in chapter 3 helping to inform the case evaluations of California, Vermont, and Georgia in later chapters. Appendix A includes more in-depth description and information about the variables used in the analysis. Appendix B provides univariate descriptions.

Internal Political Conditions

This analysis elects to include three key political determinants including state legislative professionalism, partisan control, and political culture. As part of the broader consideration for internal determinants, political conditions shed light on the propensity for key policy makers to adoption of an innovation. State legislative professionalism was selected as a political variable because of its ability to assess a state's capacity to address new innovative ideas. As described by Squire, "Measures of legislative professionalism are intended to assess the capacity of both individual members and the organization as a whole to generate and digest information in the policymaking process" (Squire 2007). In other words, some state legislatures spend more time in session, elevate levels of resources and support from staff to research and develop policy, and receive higher compensation for service. Shipan and Volden (2006) were the first to argue that state legislature professionalism may lead to adoption of innovative policies. The underlying assumption to this prediction is that innovation may only occur when resources are freed to consider policy needs that are beyond essential. Fewer days in session, lack of staffing or resource support, and poor pay may not motivate policy makers to consider unique, novel, or innovation solutions to existing problems.

Partisan control is a second political condition that may determine how policies diffuse. This particular variable assesses how partisan politics can impact adoption and reinvention decisions. The Republican-Democrat divide is grounded in predispositions for how each part will act in any given policy domain. Organic food and agriculture has some connection to environmentalism and general health concerns. Democrats have been identified as being more sympathetic to environmental causes and policy than Republicans (see Kamieniecki 1997; Kraft 2012; McCright and Dunlap 2011; Shipan and Lowry 2001). Despite environmental concerns being weak in national elections, environmentalists often vote for Democratic candidates (Duffy 2003). In addition, addressing public health concerns like expanding health insurance coverage through government regulation is also seen as more of a Democratic ideal than Republican. Therefore, it would be expected that states where Democrats have control of the legislative and executive branches would be more likely to pursue organic regulatory legislation than in divided or Republican controlled governments.

Finally, this analysis includes state political culture measures as a third political determinant. Unlike other political ideology measures including partisan control, political culture encompasses broader conceptualization for the scope of government and moves beyond the Republican-Democrat dichotomous view of politics. Daniel Elazar, a preeminent political culture scholar, defines the phenomenon as "the particular pattern of orientation to political action in which each political system is embedded" (Elazar 1972, 78). Political culture is developed in part by the settlers of specific regions in the country, and it is reflective of the history of those people in that environment.

In Elazar's own political culture construct, there are three broadly encompassing state-level political cultures in the United States. State-level cultures indicate the general perception and relationship citizens have with government. Moralistic, individualistic, and traditionalistic cultures are the three cultures identified by Elazar (1972). Each represents a different set of beliefs regarding the scope and role of government, the range of actors participating in the political process, and the normative behavior in policy making. Moralistic cultures value government service as a public good whereas traditionalist political cultures (i.e., the South) favor a more limited role of government. Individualistic cultures are more utilitarian in nature and politics is seen as a business. A state's political culture can be predominantly one culture but contain strong strains of another culture. Under Elazar's theory, a moralistic culture would be more likely to be open to government involvement in a market and pursue innovative approaches to resolving societal problems.

Internal Socio-Economic Conditions

Wealth, capacity, and size have been consistent explanations for adopting innovative policies from a socio-economic perspective (Bailey and Rom 2004; Berry and Berry 1990, 2007; Savage 1985). For this study, per capita income, fruit sales, vegetable sales, state agricultural GDP, and the presence of organic certifiers and farming organizations are used to evaluate state socio-economic conditions.

Per capita income is included in the analysis for two reasons. First, states possessing more wealth are found to be more likely to invest in innovative policies and ideas. This is likely because citizens and policy makers have more resources and desire to pursue innovative and novel ideas. Second, it is the relation between personal wealth and organic food purchases that makes per capita wealth of particular interest in this study. Specifically, wealthier individuals are more likely to purchase organic foods (Dettmann and Dimitri 2007). Per capita income serves as an indicator for a wealthier tax base, thus potentially increasing state resources, and ability to participate in the organic market.

State-level agricultural GDP addresses the market dynamic of investment. States with larger agriculture industries within the state economy may have more resources or climate-related ability to invest or experiment with alternative production methods. Evidence from studies on European Union farmers suggests that the diffusion of organic farming practices from country to country is based on the dominant farm type such as arable or specialist animal farming (Gabriel et al. 2009; Kaufmanna et al. 2011; Kerselaers et al. 2007; Michelsen et al. 2001). Within this study, the change in state GDP resulting from agriculture almost acts as a double-edge sword. It is understood that early rifts between conventional and alternative forms of agriculture may lead to misleading results in earlier time periods of the analysis. Moreover, it is also possible that organic agriculture growth may only be minimally contributing to the overall GDP growth. Yet, it is also possible that increasing GDP likely drives more ability to invest in new and innovative practices, such as organic.

Two additional industry variables include assessment of fruit and vegetable market value sales. Fruit and vegetable sales are included in the analysis as a proxy measure for organic food production. Available data on organic food production and sales does not extend the length of the study. Organic sales and production numbers were not collected across all fifty states consistently until 2000. Therefore, a proxy-measure is used to assess the presence of organic production and sales. While organic food sales still remain as only a small percentage of total food sales in the past decade, organic fruit and vegetables historically constitute the largest percentage of organic food sales in the United States. According to OTA (2011), organic fruit and vegetable sales constitute 39.7 percent of all organic sales in 2010. While other sectors of organic certified goods have seen sales expanded, produce still remains a significant market, and consumers use produce as a factor of deciding where to shop (Klonsky 2000). As an additional consideration, fruits and vegetables are considered a specialty crop, which some states are leading producers while others are not. Specialty crop production is of interest as it signifies potential organic production activity and interest in healthier food choices. Alston and Pardey (2008) argue for more research to be directed toward specialty crop production because of the "social rate of return" to making fruit and vegetables more affordable and available to include in healthy diets. As a result, the interest in fruit and vegetable sales is intended to identify how crop-specific production may lead to more interest in organic policies in some states. It is believed that higher sales levels are more likely to increase probability of adoption and reinvention.

As a final market consideration, this study also evaluates the role of organic organizations that certify operations and unite organic producers. Grassroots and industry organizations are found to be powerful drivers of policy in previous studies to both inhibit and support new legislation. In the case of organic food and agriculture, it is predicted that the presence of an organic advocacy group increases the likelihood of a state adopting legislation. This particular hypothesis is evaluated through the case studies, as there is no valid statistical measurement of organic organizations.

Issue Salience

The public policy literature, not just the diffusion literature, indicates the significance of issue salience to agenda setting, adoption, and modification of policy over time (e.g., see Baumgartner and Jones 2009; Birkland 1998; Cobb and Elder 1983; Kingdon 2003). Focusing events, indicators, and changes in the perceptions of a policy issue all contribute to increasing attention from the public, which set the systemic agenda, and policy elites, which set the formal agenda. Organic food and agriculture legislation was likely pursued in periods of elevated attention and discussion to the practice.

Vertical Dynamics of Diffusion

The adoption of OFPA in 1990 and subsequent implementation of the NOP in 2002 are two significant considerations for if adoption or reinvention of state organic policies occurred in response to federal actions. Research on vertical diffusion find authorities at the local, state, and federal levels influence policy decisions at higher and lower levels (Allen, Pettus, and Haider-Markel 2004; Boeckelman 1992; Daley and Garand 2005; Gray 1973a; Gray, Gray, and Williams 1981; Karch 2006; Shipan and Volden 2006; Welch and Thompson 1980). Federal involvement in any policy area generally spurs action among the states to comply or account for federal supremacy.

Adoption of the OFPA and subsequent legislative amendments would likely deter first time state adopters of organic regulatory legislation. However, delayed implementation likely contributed to continual state adopters in the 1990s. Once the NOP was officially implemented, it would be expected that states would amend existing policies to accommodate federal preemption. Federal implementation aids in overcoming shortfalls of relying just on federal adoption as the only factor related to vertical diffusion; the implementation measures determine how delayed implementation leads to uncertainty and may be the true mechanism leading to state policy modification and reinvention in the diffusion process. It is expected federal implementation is likely to increase the likelihood that states will adopt legislative bills modifying organic regulatory policy. This effect would occur after the implementation of the 2002 NOP final rule.

Horizontal Dynamics of Diffusion

The policy diffusion literature identifies a number of possible approaches to understanding the pressures exerted by governmental peers. Previous studies on diffusion have found a regional explanation of diffusion powerful as neighboring states may often look toward each other for how to innovate (see Berry and Berry 1990 and 2007; Walker 1969 and 1973). This analysis elects to evaluate the horizontal dimension through a regional perspective and presence of an organic certification and grower network. The latter aspect can only be observed through a qualitative evaluation and may shed some light on if and how regional and national networks operated to establish policy. This element is also qualitatively assessed as part of internal determinants. The former is based on the USDA's Agricultural Marketing Service's seven-region designation, which is the same regional designation the National Organic Program uses to evaluate a host of agricultural production and market activities (see figure 2.1), and permits a statistical evaluation. The regions collectively identify states with ecologically similar agricultural environments and food market hubs (Barham et al. 2012; Ragland and Tropp 2009). Regional adoption patterns for organic food and agriculture may be partially dependent on local neighbors that share similar market conditions and, additionally, political environments.

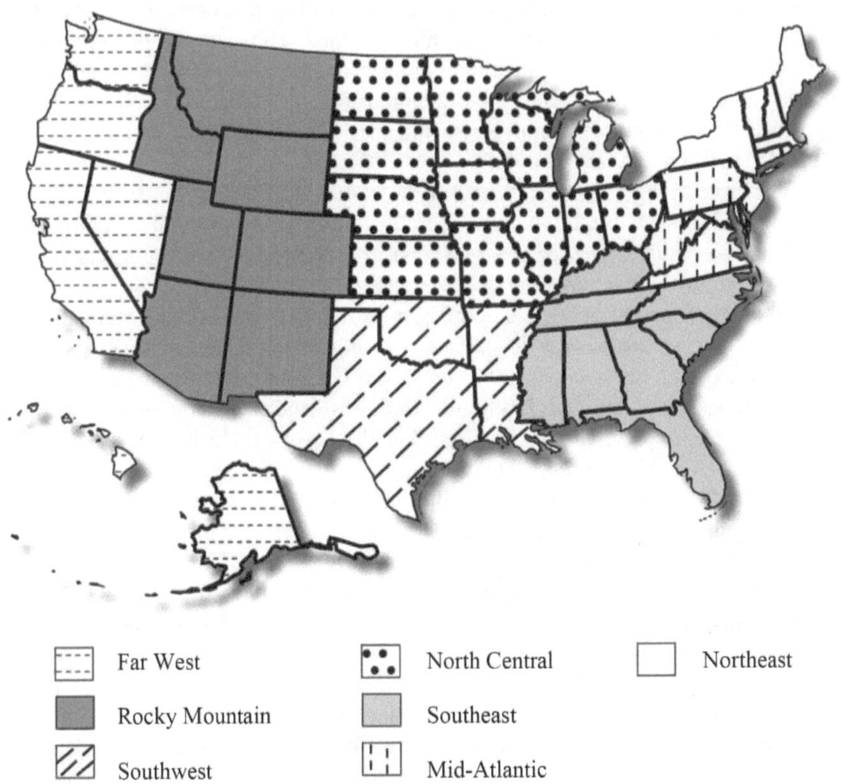

Figure 2.1. USDA AMS 7-Region Designation Map

Map created by author from data in Barham et al. 2012, Ragland and Troup 2009.

In the chapters that follow, the factors explaining the adoption and diffusion of organic regulatory legislation is explored. The policies developed and enacted by state legislatures to define organic practices and permissible labeling are the product of a political environment keenly aware of the economic and political consequences associated with enactment. The information and models presented in chapter 3 provide a general overview of why and how policies diffused. The cases presented in chapters 4–6 detail how organic regulations emerged and evolved in three different cases.

NOTES

1. According to Lejano, Ingram, and Ingram (2013) there are several alternative farming networks that have operated since the 1940s including eco-agriculture, biodynamic, and organic farming. Each network subscribes to a particular set of organic techniques and is influential in a number of movements for organics. For additional information about the rise of specific organic farming practices see Lockeretz's (2007) "What Explains the Rise of Organic Farming?" However, despite the identified differences in alternative agricultural production, this analysis will take a broadly conceived notion of the organic agriculture industry.

2. Key figures within this movement include Edward Faulkner, Jerome Rodale, Louis Bromfield and Aldo Leopold.

3. The iron triangle concept refers to the stable relationship between congressional committees, an executive branch administrative agency, and interest groups to maintain a particular policy or program. The iron triangle is believed to result in a monopoly of power in a particular policy area.

4. Butz is credited as being one of the best-known and controversial secretaries of agriculture. His remark on organic agriculture is only one of many controversial statements made during his five-year term (see Goldstein 2008). His service as secretary of agriculture came to an end during the 1976 presidential election after he made a series of insensitive racial remarks regarding blacks' "sexual, dress and bathroom preferences" (Associated Press 1976).

5. Oregon Tilth began as Regional Tilth in 1974 with chapters in Oregon, Washington, Idaho, and Northern California (Oregon Tilth N.A.). The Willamette Valley chapter became Oregon Tilth in 1986.

6. CCOF emerged from the California Organic Farmers Association (COFA), which formed in 1972 (CCOF 1988). COFA was primarily focused on marketing instead of certification.

7. IFOAM formed in 1972 in Versailles, France. In 1980, IFOAM passed the International Basic Standards for Organic Production and Processing, which the FAO considers as "the private sector's equivalent to the Codex Alimentarius guidelines" (FAO, n.d.).

8. The Codex Alimentarius Commission was created in 1963 as a joint effort between the U.N. Food and Agriculture Organization (FAO) and the World Health Or-

ganization (WHO) to harmonize international food standards, guidelines, and codes of practice. In 1999, the commission first adopted the *Guidelines for Production, Processing, Labeling and Marketing of Organically Produced Foods*.

9. In 2003, a joint venture was undertaken by the FAO, IFOAM, and the United Nations Conference on Trade and Development to reconcile the differences between two sets of guidelines in light of many standards, regulations, and certification requirements that created trade barriers. The joint venture resulted in the development of the Common Objectives and Requirements of Organic Standards (COROS) in 2010, which seeks to provide equivalence assessments of organic standards and regulations. For more information see http://www.goma-organic.org/news-from-goma/common-objectives-and-requirements-for-organic-standards-coros-makes-its-debut/.

10. After official implementation in 2002, Arthur Harvey filed a civil lawsuit with nine counts challenging the NOP as not in compliance with OFPA. In January 2005, The U.S. Court of Appeals for the First Circuit ruled in favor of Harvey on three issues, which required the USDA NOP to revise existing rules in 2006. For more detailed information see http://www.ams.usda.gov/AMSv1.0/getfile?dDocName=STELDEV3013543.

11. States covered in the AMA program include Connecticut, Delaware, Hawaii, Maine, Maryland, Massachusetts, Nevada, New Hampshire, New Jersey, New York, Pennsylvania, Rhode Island, Utah, Vermont, West Virginia, and Wyoming.

Chapter Three

Organic Policy in the States

New York was the first state to adopt organic legislation permitting regulations for the organic food and agriculture market. In the years that followed, only a few additional state legislatures pursued organic policy. Yet, by the late 1980s, a critical mass followed where states rapidly sought regulatory policies of their own. Fewer additional states continued to pursue organic policies of their own after the Congress adopted the national Organic Foods Productions Act of 1990, despite federal supremacy. What caused this particular pattern of diffusion? The diffusion of organic regulatory legislation, one dimension of the overall policy, is the result of internal and external determinants as expected, but timing is a key condition for understanding determinants of diffusion. Innovative leaders of organic regulatory policy were influenced by a different set of conditions than laggard adopters, as environmental conditions changed politically and economically over the course of thirty-five years. The results presented in this chapter provide a generalized snapshot of how diffusion progressed among U.S. states and the determinants significant to diffusion. In what follows, a broad pictorial and statistical overview of conditions of diffusion is presented. It is followed by a more narrowed assessment of diffusion eras, a discussion of mechanisms of diffusion, and an overview of policy termination.

A COMPREHENSIVE PICTURE OF ORGANIC DIFFUSION

From 1976–2010 a total of thirty-eight states passed legislation regarding the regulation of organic food and agriculture; thirty states would modify existing state statutes through one of the 107 policy amending bills adopted from 1980–2010. This analysis only assesses legislation that substantially created, changed, or terminated statutory language regarding regulations or the permissible creation of regulations for the organic food and agriculture market. This includes, but is not limited to, the creation of councils, advisory boards, authorization to state agencies or departments to create rules, and actual regulations regarding appropriate labeling or agricultural practices, production, processing, distribution or sale of organic items. This analysis examines any adopted legislation (i.e., any adoption), first time or initial legislation adoption (i.e., initial adoption), and the adoption of legislation that amends, reinvents, or terminates existing statutory language (i.e., amending adoption).

The speed of adoption conforms to the s-shaped curve of diffusion (see figure 3.1). Most states adopted organic legislation for the first time from 1985–1990 with minimal growth before and after this period. The speed of diffusion is neither rapid nor incrementally slow; it does, however, show the process extending across a lengthy time period of thirty-five years. The pattern seen in figure 3.1 aids in focusing the analysis into two key time periods beyond a comprehensive perspective as it identifies key groupings of state adopters.

Pre-1990 adoptions include a pack of leading innovators and middle-of-the-pack adopters. Leading innovators include New York, California, Connecticut, and Maine, all of which adopted prior to 1980. Spatially, two regions, the Northeast and the Far West, act as regional leaders with New York and

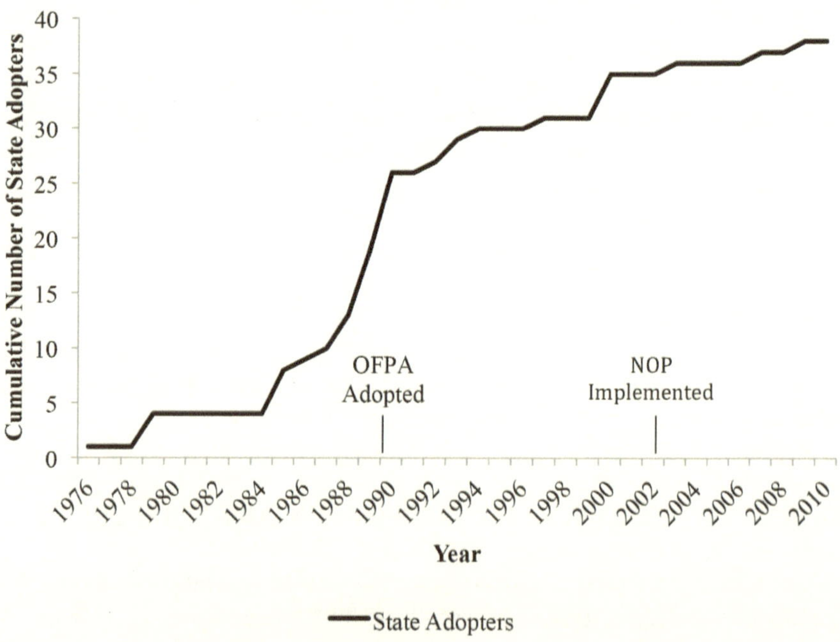

Figure 3.1. Cumulative Total of State Adopters

Created by the author with data compiled from state legislative databases and law libraries.

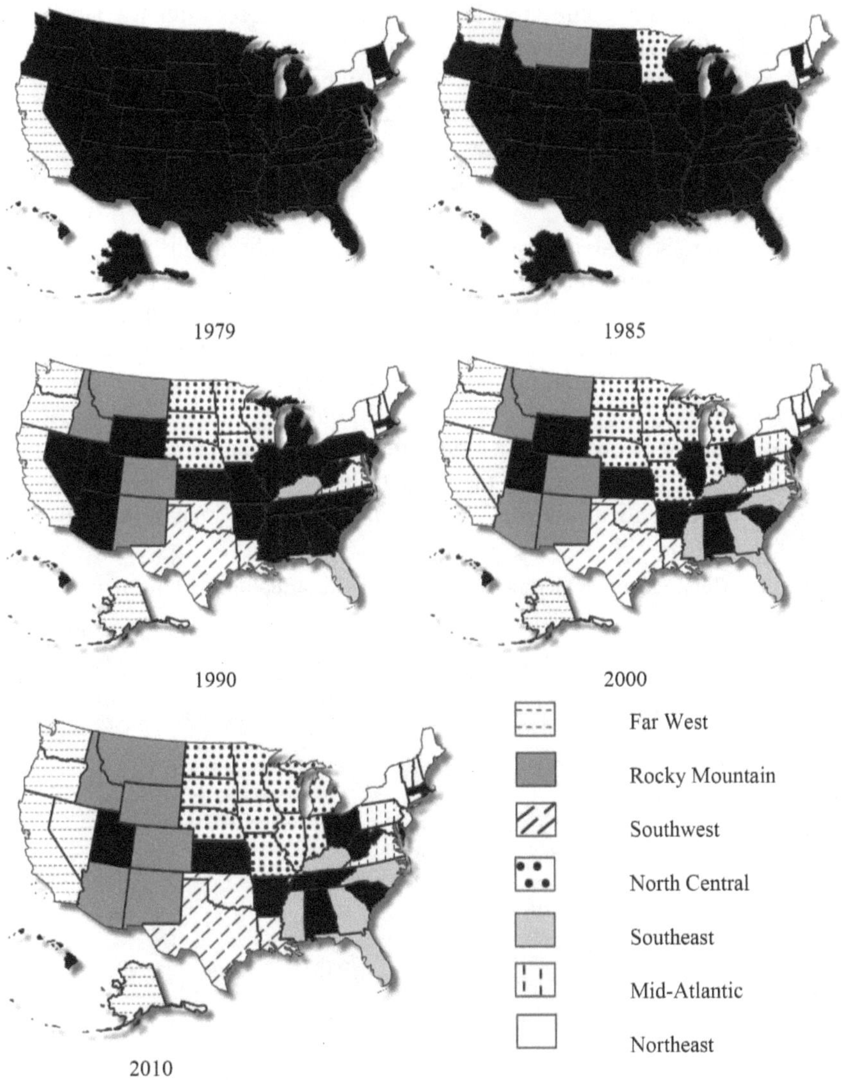

Figure 3.2. Diffusion by Region

Data compiled by the author from state legislative databases and law libraries; maps created by author from data in Barham et al. 2012, Ragland and Troup 2009.

California serving as each region's pacesetter (see figure 3.2). Middle-of-the-pack adopters include twenty-two states that adopted legislation from 1985–1990. During this time period, states in the central region of the United States began to follow the regional cues set by the Far West and Northeast. The Southeast would remain a holdout region until 1990, with only 25 percent of the region (i.e., Kentucky and Florida) adopting legislation in 1990.

Laggard adopters come later in the diffusion process after the Congress adopted OFPA. The five states that adopted from 1992–1997 are difficult to classify as lagging adopters. It was during this period that uncertainty surrounded the implementation of the National Organic Program. The same cannot be said for the seven states adopting in 2000 or after, which are true innovative laggards in this policy domain. The laggards are not confined to one geographical region.

From a comprehensive perspective, a range of conditions influenced states' decisions to adopt organic regulatory legislation from 1976–2010 (see table 3.1). Any adoption, whether initial or amending existing statutes, are influenced by both internal and external conditions. If narrowed by type of adoption (e.g., initial or amending), the nuanced variations in type of adoption help in explaining the overall assessment. Initial state adoptions from 1976–2010 were more likely to occur under the following conditions: Democrats controlled state government (a factor that almost doubles the probability of adoption); the political culture was moralistic-leaning in nature; issue sa-

Table 3.1. Comprehensive Picture of State Legislative Adoptions (1976–2010)

	Any Adoption		*Initial Adoptions*		*Amending Adoptions*	
Variable	*Odds Ratio*	*Std. Error*	*Odds Ratio*	*Std. Error*	*Odds Ratio*	*Std. Error*
Partisan Control	1.39*	.218	1.75†	.566	1.18	.208
State Legislative Professionalism	.159†	.1565	.485	1.05	.104*	.115
Political Culture	.869**	.034	.756**	.062	.855**	.037
Per Capita Income	.999	.000	1.00	.000	.999	.000
Log Agriculture GDP	1.39**	.177	1.02	.250	.139*	.210
Log Ag. Sales from Fruit	.958	.064	.953	.134	.996	.076
Log Ag. Sales from Veggies	1.34**	.125	1.05	.206	1.45**	.154
Issue Salience	1.26	.197	3.22**	.801	1.49**	.218
Federal Adoption	1.85**	.371	2.09†	.861	1.82**	.413
Federal Implementation	.488**	.131	.130**	.096	.605†	.105
Regional Adopt Percentage[+]	2.42	1.53	.004**	.096		
	Log likelihood = −429.769 Prob>chi2 = .0000		Log likelihood = −147.74 Prob>chi2 = .0000		Log likelihood = −336.23 Prob>chi2 = .0000	

†Significant at 10 percent level, *Significant at 5 percent level, **Significant at 1 percent level

[+]Regional adoption percentage was excluded from amending adoption model to prevent false positive results.

Source: Data compiled by the author; see appendix A.

lience increased; and years in which Congress adopted organic bills. Federal implementation and regional pressures are significant but have minimal effects on adoption that decrease the probability of adoption.

Comparatively, amending adoptions were also influenced by federal adoption, federal implementation, and issue salience, but increasing agricultural marketing activities, including vegetable sales, and professionalized state legislatures also matter for reinvention. This suggests that reassessment by legislatures was more likely to occur if the state had a vested interest both economically and politically to do so. Even modifications in response to federal adoption and implementation are political reactions by legislators to ensure federal supremacy in this policy domain.

PRE-1990 ADOPTIONS: INNOVATIVE ORGANIC LEADERS AND FOLLOWERS

State legislative action prior to the passage of OFPA in 1990 covers a fourteen-year time period from 1976–1989. Two time frames are used to analyze the initial adoption of organic bills during this period.[1] The first time frame examines all fourteen years prior to 1990 while the second time frame focuses on the initial adoptions occurring during the critical mass from 1985–1990, when the majority of states adopted.

Two key conditions, both of which appear significant in the comprehensive outputs, have a significant impact on early adopters of organic regulatory legislation (see tables 3.2 and 3.3).[2] Political culture has a positive effect

Table 3.2. Pre-1990 State Legislative Adoptions

	Any Adoption		*Initial Adoptions*	
Variable	*Odds Ratio*	*Std. Error*	*Odds Ratio*	*Std. Error*
Partisan Control	1.91*	.734	1.33	.634
State Legislative Professionalism	.030	.074	.480	1.41
Political Culture	.651**	.079	.685**	.093
Per Capita Income	1.00	.000	1.00	.000
Log Agriculture GDP	1.42	.453	1.17	.430
Log Ag. Sales from Fruit	1.11	.206	.978	.204
Log Ag. Sales from Veggies	.896	.245	.937	.287
Issue Salience	2.31**	.545	3.07**	.874
Regional Adopt Percentage	3.44	4.95	.161	.291
	Log likelihood = –86.40 Prob>chi2 =.0000		Log likelihood = –64.84 Prob>chi2 = .0014	

†Significant at 10 percent level, *Significant at 5 percent level, **Significant at 1 percent level

Source: Data compiled by the author; see appendix A.

Table 3.3. State Adoptions during Critical Mass (1985–1990)

	Any Adoption		*Initial Adoptions*	
Variable	*Odds Ratio*	*Std. Error*	*Odds Ratio*	*Std. Error*
Partisan Control	1.21	.626	1.07	.603
State Legislative Professionalism	.030	.112	.553	1.98
Political Culture	.661**	.102	.652**	.100
Per Capita Income	.999	.000	.999	.000
Log Agriculture GDP	1.99	.929	1.35	.617
Log Ag. Sales from Fruit	.938	.231	.898	.222
Log Ag. Sales from Veggies	.949	.373	.938	.352
Issue Salience	2.24**	.668	2.65**	.903
Regional Adopt Percentage	.460	.905	.026†	.052
	Log likelihood = –43.6 Prob>chi2 =.0052		Log likelihood = –43.97 Prob>chi2 = .0221	

†Significant at 10 percent level, *Significant at 5 percent level, **Significant at 1 percent level
Federal adoption was not intentionally omitted from both models.

Source: Data compiled by the author; see appendix A.

with moralistic-leaning cultures being more likely to adopt organic legislation. For middle-of-the-pack adopters (e.g., states adopting from 1985–1990), each increment closer to a pure moralistic political culture increases the probability of adoption by 35 percent. When including innovative leaders (e.g., those before 1985), the percentage slightly drops to 32 percent. These results suggest that early adopters, including both innovator leaders and those adopting during a critical mass, likely possess political cultures that are open to the idea of regulating sustainable agriculture. Moralistic cultures are described as being more trusting of government and seeking policies that are for the public good. Organic agriculture policies could be seen as protecting an alternative production method that seeks to improve public health and the quality of the environment. Identifying which products are organic compared to conventional also corrects market asymmetries. This perspective is not necessarily held by individualistic or traditionalist cultures. Traditionalist cultures generally mistrust a large government role in society whereas individualistic strains have an emphasis on individual rights with a healthy dose of government skepticism.

Issue salience is the second key condition to explaining early and middle-of-the-pack legislative action. State legislators were three times more likely to adopt organic legislation as attention to the issue elevated. Four articles appeared in the *New York Times* in 1973. By 1989, this figure jumps to twenty-five articles representing a 525 percent increase in eighteen years (see figure 3.3). However, attention to the issue would drop in 1990 but would continue to

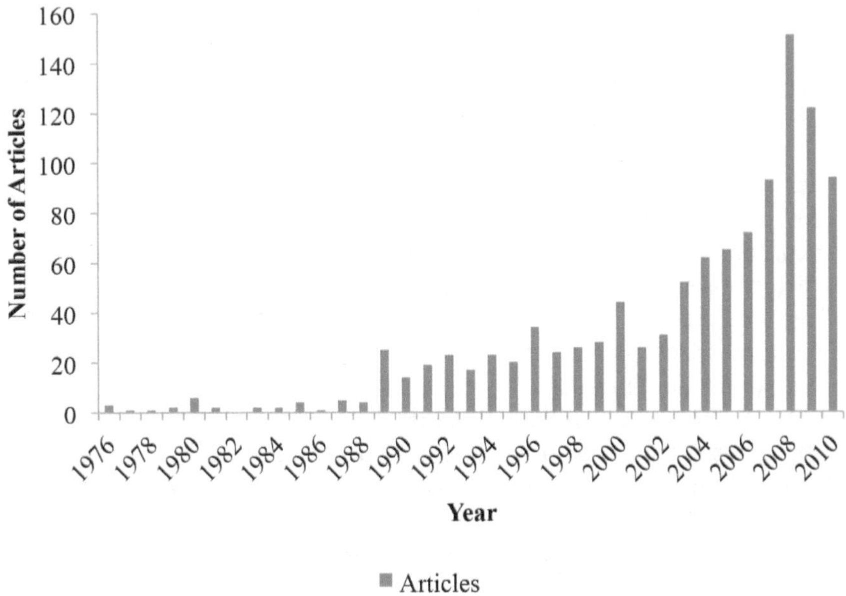

Figure 3.3. Organic Food and Agriculture Issue Salience by Annual Number of Articles in the *New York Times*

Created by the author with data compiled from multiple sources; see appendix A.

grow until 2008 before declining again. For pre-1990 adopters, higher levels of attention to the issue mattered for policy makers. The same cannot be said for later adopters. The post-OFPA era had even higher number of articles dedicated to organic food and agricultural concerns but this did not translate into a statistically significant effect for states that adopted on or after 1990.

In addition to issue salience and political culture, two other variables appear as significant to pre-1990 organic bills. First, regional effects have a minimal effect on pressuring states to adopt from 1985–1990. States adopting during the critical mass period may have been looking to their neighbors for behavioral cues, but the effect is relatively limited and suggests a pressure valve effect that suggests pressure relief to adopt. Second, partisan control indicates that Democratic-controlled state governments are more likely to adopt bills than governments under divided or Republican control. The effect, however, is limited to any adoption during the period and not just initial adoptions. If looking at the comprehensive models, partisan considerations are not a significant explanatory variable for amending adoptions either. This may suggest that partisanship played a role in debates but did not matter specifically for initial and amending adoptions isolation. Instead, the overall effect is cumulative in nature.

IN AN OFPA ERA—LAGGARDS OF ORGANIC INNOVATION

The environment of organic policy making drastically changed in 1990. The adoption of the Organic Foods Production Act of 1990 marked an era of federal dominance in developing and overseeing regulatory policy. The role states had served prior to 1990 was one of creating and shaping what was and wasn't organic. The role after 1990 was bound to considerations of federal action. Initial adoptions on or after 1990 were not influenced by anything in particular except, perhaps, regional considerations (see table 3.4). The effect of regional pressure is minimal and suggests a pressure valve effect, which suggest other factors not statistically assessed may have mattered for the twenty-five states that adopted from 1990–2010.

What becomes clear in the OFPA era is some indication of how and why states reinvented policies in the latter half of the diffusion cycle. Strong explanatory factors include increasing vegetable sales, increasing state agricul-

Table 3.4. Post-OFPA State Adoptions (1990–2010)

	Any Adoption		*Initial Adoptions*		*Amending Adoptions*	
Variable	*Odds Ratio*	*Std. Error*	*Odds Ratio*	*Std. Error*	*Odds Ratio*	*Std. Error*
Partisan Control	1.39†	.247	1.95	.877	1.21	.230
State Legislative Professionalism	.134†	.153	.361	1.25	.141†	.170
Political Culture	.927	.043	.833	.108	.892*	.042
Per Capita Income	1.00	.000	1.00	.000	1.00	.000
Log Agriculture GDP	1.44*	.218	.839	.307	1.55**	.259
Log Ag. Sales from Fruit	.900	.068	.882	.177	.921	.075
Log Ag. Sales from Veggies	1.60**	.171	1.38	.390	1.58**	.179
Issue Salience	.391**	.352	2.45	2.06	.385**	.128
Federal Adoption	1.58*	.352	1.81	.970	1.28	.312
Federal Implementation	1.60	.620	.250	.278	2.33*	.991
Regional Adopt Percentage⁺	1.53	1.19	.001**	.002		
	Log likelihood = –317.23 Prob>chi2 = .0000		Log likelihood = –75.301 Prob>chi2 = .0001		Log likelihood = –276.371 Prob>chi2 = .0000	

†Significant at 10 percent level, *Significant at 5 percent level, **Significant at 1 percent level
⁺Regional adoption percentage was excluded from amending adoption model to prevent false positive results.

Source: Data compiled by the author; see appendix A.

tural GDP, and federal implementation. An increase in vegetable sales with an increasing agricultural GDP increased the probability of legislative change by about 1.5 times while federal implementation increased the probability by 1.8 times. State legislative professionalism, political culture, and issue salience also matter albeit in not the same strength or significance. The findings generally conform to the expectations outlined for this evaluation (see chapter 2) and show key differences between environmental conditions significant to first time adoptions versus states seeking to reinvent organic statutory language.

Later adopters, including those laggards in the 2000s, were operating in an environment where a majority of states had already pursued organic legislation and the federal government became involved to correct the patchwork state policies. Federal adoption did not spur reinvention in the same capacity that it did not have a positive or negative effect on adoption of initial state legislation. One potential explanation for this outcome is that states were cautious of the lag in establishing the National Organic Standards Board and development of a final rule. However, states were willing to reassess existing policies once federal law was to be implemented.

In addition to the vertical pressures of diffusion, states with internal drivers likely to permit consideration of agricultural policies were more likely to reinvent. The increase in economic activity in both the broader agricultural market could exert more pressure on a state to ensure agricultural laws as representative and supportive of that market. This is particularly true in a post-OFPA era where broader corporate interest in the organic market materialized. Correspondingly, states with a smaller agricultural market or whose producers specialize in non-specialty crop production may be less likely to readjust a policy that only concerns a small segment of the state's population. Compared to fruit sales, vegetable sales could be significant because vegetables are more often annual rather than perennial. In addition, fruit production is geographically limited compared to vegetable production. Nevertheless, all indicators point to states with more capability and desire to adjust policies as being more likely to modify and fine-tune existing policy.

The Legislative Death of Organic Policy

Reinvention or policy change may sometimes result in the death of existing statutory code. After the adoption of OFPA and implementation of the NOP, it could be expected that some state legislatures seek to terminate statutory language regulating or authorizing the development of organic rules. In essence, state legislatures may opt to remove language from state statutes that is no longer needed or wanted in order to comply with federal law.

Death is an event that has not happened frequently with organic policy. From 2001–2010, nine state legislatures terminated or all-but-terminated organic statutory language including Indiana, Kentucky, Maine, New York, North Dakota, Oregon, Pennsylvania, Vermont, and Virginia.[3] If accounting for administrative rules, two states, Ohio and Massachusetts, have fully terminated their policy. Most legislative terminations occurred in the years preceding and following the NOP final rule implementation in 2002 (see figure 3.4).[4] The statistical results of post-OFPA model show amending legislative adoptions as positively influenced by federal implementation. This same influence may have carried over and inspired termination of legislative statutes as well. There is a cluster of five states that adopted termination legislation in the years surrounding NOP implementation. However, there were also four states that waited to remove statutory language until the latter half of the decade. In the case of New York, which terminated language pertaining to an advisory board in 2010, the termination was an effort to clean up "outdated laws" (New York Agriculture Specialist, Personal Email Communications, June 8, 2013). It is unclear if other states during this time period, including Kentucky, Oregon, and Pennsylvania, were operating under similar premises. Nevertheless, this pattern of state organic policy death, from a legislative perspective, suggests that states still sought to provide regulatory policy guidance in the marketplace long after the adoption of the OFPA in 1990 and NOP implementation in 2002. It leads to questions surrounding the validity and long-term viability of the national organic food regulations.

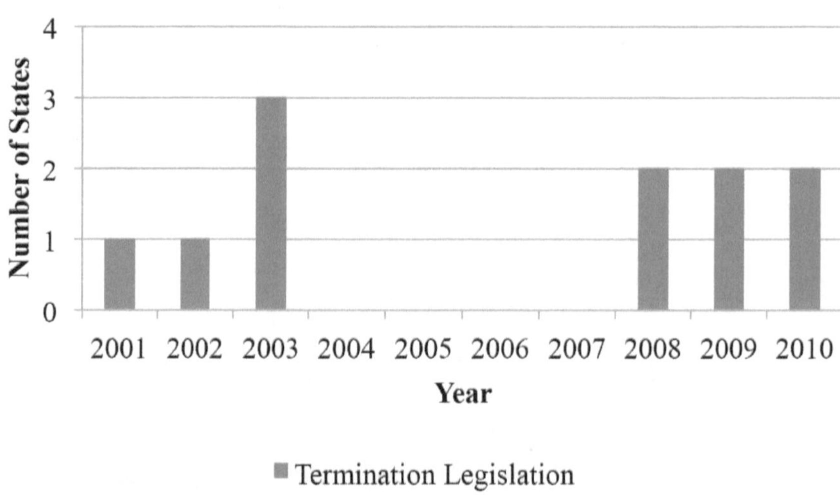

Figure 3.4. State Legislative Termination by Year

Created by the author with data compiled from state legislative databases and law libraries.

HOW DID ORGANIC POLICY DIFFUSE? MECHANISMS OF DIFFUSION AND OTHER QUESTIONS

The diffusion of organic food and agriculture legislation is influenced by a host of internal, horizontal, and vertical determinants. While significant to understanding the diffusion process, it is unknown as to how exactly the policy spread. In the case of organic legislation, no two states adopted identical initial legislation or amended state statutory language in identical manners. This is not to say that imitation as a pathway did not occur, but a majority of states varied in complexity, content, and scope, even accounting for reinvention over time. The closest case of pure imitation was among North Dakota (adopting in 1987) and South Dakota (adopting in 1988). The *only* difference between the two bills is the inclusion of a definition for a secretary in the South Dakota bill. The remaining state adopters customized legislation to suit individual needs and kept to a general basis on similar content such as providing a definition for organic production and identifying acceptable practices and labeling. For implementation design, some states outlined specific regulatory rules in legislation while others created organic advisory boards, creating third-party certification agreements, or delegated rules development and certification responsibilities to state agencies. In general, later initial adopters had policies that were no more complex than that of leaders. Some of this may be the result of depending on boards and agencies to develop specific rules and regulations.

The policy content of both initial and amending legislative bills do suggest two mechanisms of diffusion, learning and competition, which drove the content and scope of organic legislative bills perhaps more so than the complexity. Initial policy content and complexity varies among the states. Some states, like California, had incredibly detailed legislation for the time that was intended to exclude political tinkering by California's Department of Public Health and Department of Agriculture (see chapter 4 for more detail). The law also made enforcement by the agencies optional. Washington, while not as detailed, also made enforcement optional in 1985. The language used in California's initial bill is borrowed from Oregon's 1973 administrative rules, perhaps explaining the depth of California's bill.

Comparatively, New York and Maine's initial legislative adoptions were not as detailed despite Massachusetts establishing rules in 1978 and northeastern organic farmers interacting with west coast organic associations such as Oregon Tilth and California Certified Organic Farmers. New York's 1976 bill merely requested an investigation into the need and options for developing organic rules by the New York Department of Agriculture and Markets. Rule development was left as optional. Maine's 1979 bill outlined key terms, similar to California, Oregon, Massachusetts, and Washington, but was not as extensively detailed nor contained voluntary enforcement measures.

Later adopters also varied in content and scope. For example, Wisconsin's 1987 Act 278 is a one-paragraph authorization for the Department of Agriculture, Trade, and Consumer Protection to promulgate rules. In 1989, Vermont's bill sought to not only authorize the development of organic rules by an advisory council but also invest in sustainable agricultural research and practicum in universities, a policy cousin of organic regulation (see chapter 5). This was a collective effort to permit certifying agencies to make better judgments of permissible practices rather than succumb to potentially perverted-involvement by government and conventional agriculture interests. Kentucky's 1990 bill was more interested in general labeling of state food products and included organic labeling as more of an afterthought to address an emerging concern (Personal Communications, Former Kentucky State Legislator, 6/18/15). By 2000, Georgia's organic legislation covered the basics as a "just in case" measure if the National Organic Program failed to adopt final rules (Personal Communication, Former Georgia Administrator, 10/29/2013). At the time there was concern on whether the NOP's proposed and final rules would succeed (see chapter 6). One of the last adopters, Wyoming in 2009, had legislation that exclusively focused on organic fertilizers and the effect on labeling. California also reinvented its policy the same year to address organic fertilizers but was not similar to Wyoming's bill. State action was more than likely motivated by discrepancies in federal organic policy that was not fully addressed until 2010.

Accounting for the dynamic differences in content, scope, and complexity the story of diffusion can only partially be told from a statistical standpoint. Explaining how variation in policy occurs and illuminating how specific determinant matter for legislative adoption decisions can only be achieved through further in-depth investigation. There are other determinants that likely may explain how the policy diffused and not just the conditions that permitted adoption. The three case studies presented in this book highlight varying conditions present in state organic policy adopters. California is a case of a leading innovator. Located in the Far West region, California initially adopted legislation in 1979. Since this time the state has amended its policy to respond to emerging concerns in the market, to accommodate federal preemption, and also to challenge and drive national standards for organic production. Vermont is a case that represents middle-of-the-pack adopters. Located in the Northeast, Vermont's legislature spent two years seeking to adopt a bill in 1990 that would create an organic advisory board that would challenge the consumer-perspective federal law. Finally, the case of Georgia represents a laggard adopter and laggard region. Adopting legislation for the first time in 2000, this Southeastern state is one of the last states in the country to pursue organic regulations. The cases detail how certain mechanisms, like the presence of organic farming and certification networks, were significant drivers for developing and adopting organic legislation. In addition, the determinants of adoption and diffusion also show why there is variation in content, complexity, and scope of organic policies.

NOTES

1. Models evaluating amending adoptions are not included as frequency of occurrence is low. Only ten bills modifying state statutes were adopted from 1980–1989.
2. Federal adoption was included in the 1985–1990 model but omitted.
3. In the case of partial termination or all-but-terminated, state legislatures opted to remove all statutory sections from existence but opted to keep a few lines authorizing state agencies to monitor or enforce organic laws and programs, both state and federal.
4. Kentucky terminated statutory language over the course of two bills. Partial termination occurred in 2002 and final termination was complete in 2008. However, authorization for a state certification program remained.

Chapter Four

Organic in California

A Leading Innovator

The story of organic agriculture in California spans over four decades. While not the first state to adopt organic food and agriculture legislation, California is a leading innovator and has pushed the boundaries for organic policy development. The state played an instrumental role in moving organic policy forward on a national and international scale, and it is the only state to be approved to operate its own state organic program. Since the initial adoption in 1979, California has amended its Health and Safety code and Food and Agricultural code seventeen times from 1982–2010. The number of amending adoptions, which reinvents the state policy, far exceeds other states.

The case of California illuminates the significance of organic producers and certifying networks to the development of organic regulations in the state. The California Certified Organic Farmers organization was instrumental for the emergence of state law and captures how the political culture of the state influenced the policy design. However, only part of the story can be told from the perspective of third-party certifiers and organic advocacy groups. Other conditions also spurred adoption of legislation and reinvention of the policy including partisan control, issue salience, regional and national interactions with other state organic organizations, and the interplay between California's laws with the federal law.

This chapter begins with a brief overview of politics and the agricultural industry in California. Details of California's transition to specialty crops are included. Next, an overview of the emergence and growth of the organic industry is discussed. Then, California's organic legislative adoptions are detailed. A total of eighteen bills were adopted from 1979–2010. The history of California's organic policy is divided into three eras marked by the Organic Foods Act of 1979, the California Organic Foods Act of 1990, and the California Organic

Products Act of 2003. Finally, this chapter concludes with an evaluation of significant factors of adoption and an overview of the case study's contribution to the knowledge of organic policy and policy diffusion.

CALIFORNIA POLITICAL AND SOCIO-ECONOMIC CONDITIONS

The Golden State is a unique case study in any policy evaluation. California is politically and economically distinct compared to most other states in the United States. The economy is one of the largest in the world. The state is home to a wide range of industries that promote economic development and sustain economic performance. Politically, the state is also varied with regional hubs of conservatism and liberalism. This may contribute to divided partisanship and a political culture that is reflective of both trust and skepticism over the purpose and role of government in society.

Politics in California

The political environment in California is complex, but in more recent years there is favoritism toward Democratic candidates rather than Republicans. In presidential elections from 2000–2012, Democrats have fared well. However, the state hasn't always swayed Democrat and there are clear regional differences along a Republican-Democrat divide (see appendix C). Since 2000, the coastal areas of California have majority support for Democratic candidates whereas interior regions voted predominantly Republican. On average, Democratic-leaning areas are wealthier and have higher population densities (see appendix C).

California shares similarities with other states regarding partisan control of state government (see appendix C). Most states within the West region have been controlled by either Democrats or under split control. California shares a similar pattern with Nevada and Oregon with the government being Democrat versus split control over time. For much of the past three decades, California and Nevada have been under split government control, where neither party controls state government. California is also similar to Nevada with regard to political culture. California's political culture is identified as moralistic with a strong individualistic strain (Elazar 1994, 284), which correlates to a 3.55 ranking on Sharkansky's numeric scale (1969). California's political culture, therefore, has a concern for the public good but also strongly supports the rights of the individual. In general, the broad designation of California's political culture demonstrates trust in government to address public concerns while there is an undertone of deep mistrust of government, as well. Comparatively, Nevada is the only other state in the region with a similar political culture with a 3.66 political culture ranking on Sharkansky's scale. Oregon and Washington have political cultures ranking more purely moralistic, whereas Alaska has an individualistic culture and Hawaii has a political culture closer to being traditionalist.

California differs from other states in the West with regard to the level of state legislature professionalism. California's state legislature professionalism ranges from .626 as a peak in 2003 to a low of .526 in 1979. While California's professionalism score follows a similar fluctuation pattern over time, Alaska, Hawaii, Nevada, Oregon, and Washington all have state legislature professionalism scores lower than California by at least .200. Nevada, with which California shares other similar political environmental conditions, maintains the lowest regional level of state legislature professionalism with a peak of .163 in 2010 and a low of .130 in 1979. California's higher level of state legislative professionalism indicates the state legislature spends more days in session, has more resources, has a larger staff, and is considered full-time. In sum, the political conditions in California are similar to other states in the region. Yet, California's political conditions are distinctive with the state being more professionalized than other states.

Agriculture in California

Over the past century, California has consistently ranked as one of the top or the top agricultural producer in the United States (Ball, Wang, and Nehring 2013). While agriculture barely contributes to the state's current overall GDP,[1] California's agricultural industry plays a significant role in the United States and global agricultural markets. The Golden State not only has the largest agricultural production and sales figures in the United States, but it also has the most diversified set of crop production, including nuts, fruits, vegetables, horticulture, dairy, poultry, and cattle products (Olmstead, and Rhode 2003). California is the leading producer for several products, contributing to 99 percent or more of U.S. production totals, including almonds, artichokes, dates, figs, raisins, kiwis, olives, pistachios, prunes, and walnuts among other commodity and specialty crops (Sumner, Bervejillo, and Kuminoff 2003, 61). On the global market, top export destinations for California agricultural commodities include Canada, China, the European Union, India, Japan, and the United Arab Emirates (Agricultural Issues Center, n.d.). Total value of the California's 2014 exports is $21.59 billion with almond exports being the foremost export valued at $4.5 billion (Agricultural Issues Center, n.d.). California's share of the total 2014 U.S. agricultural exports was 14.3 percent.

The agriculture industry is a vital part of California to the extent it can be argued, "California's politics, economy, culture, society, environment and technology are inextricably tied to agriculture" (University of California-Berkeley Library [UCB-Library], n.d.). California's agricultural industry benefits from a two-season cycle of rainy winters and dry summers that permit year-round production (UCB-Library, n.d.). As such, California's Mediterranean climate leads to diversified production. There are several ways to demonstrate the regional variance of agricultural production in California. State agencies and universities have created several approaches to identifying agricultural regions within the state. State agricultural region maps have ranged from including four to eight different regional designations based on production areas largely based on different resources, land uses, and climatic patterns in each county (EDD 2010; Johnston 2003). Figure 4.1 demonstrates one example of regional agriculture production. The Central Coast, Sacramento Valley, and San Joaquin Valley are key agricultural production centers largely responsible for most of the state's agricultural production (Johnston 2003, 34–37). However, none of the regions would be as productive today had it not been for regional settlement patterns and the use of agricultural technology.

It was only after the 1760s that California could be considered a "mecca for agriculture" (UCB-Library, n.d.). Prior to early settlers in the region, the California region lacked native crop species. Settlers in the region, predominantly Spanish missionaries, were the first to introduce non-native crops including wheat, grapes, nuts, and cotton. For much of the eighteenth and nineteenth centuries, California agriculture was predominantly large-scale wheat farms (Olmstead and Rhode 2003, 2–3). After decades of poor agricultural management practices (e.g., lack of crop rotation or deep plowing), wheat farming became a fruitless endeavor with yields and crop quality sharply declining. According to Olmstead and Rhode (2003, 3–5), the farm economy shifted from large-scale operations to small-scale specialty crop production. The rapid growth of irrigated land at the beginning of the twentieth century coincided with the sharp increase in specialty crop production. Emerging cooperatives produced many of the specialty crops. For example, Sunkist was a label developed by the California Fruit Growers Exchange established in 1905 (Olmstead and Rhode, 23). Olmstead and Rhode (2003, 24) claim that the success and power of the California co-op model was not matched anywhere else in the country due to the state's dominance in producing many specialty crops like almonds and grapes.

Since the end of World War II, specialty crops are not the only area for rapid agricultural production growth. The cotton and livestock industries also grew substantially (Olmstead and Rhode 2003, 7–11). In part, the large growth of these three industry sectors can be attributed to the creation of cooperatives, increased prevalence of mechanized farms, which includes irrigation, and use

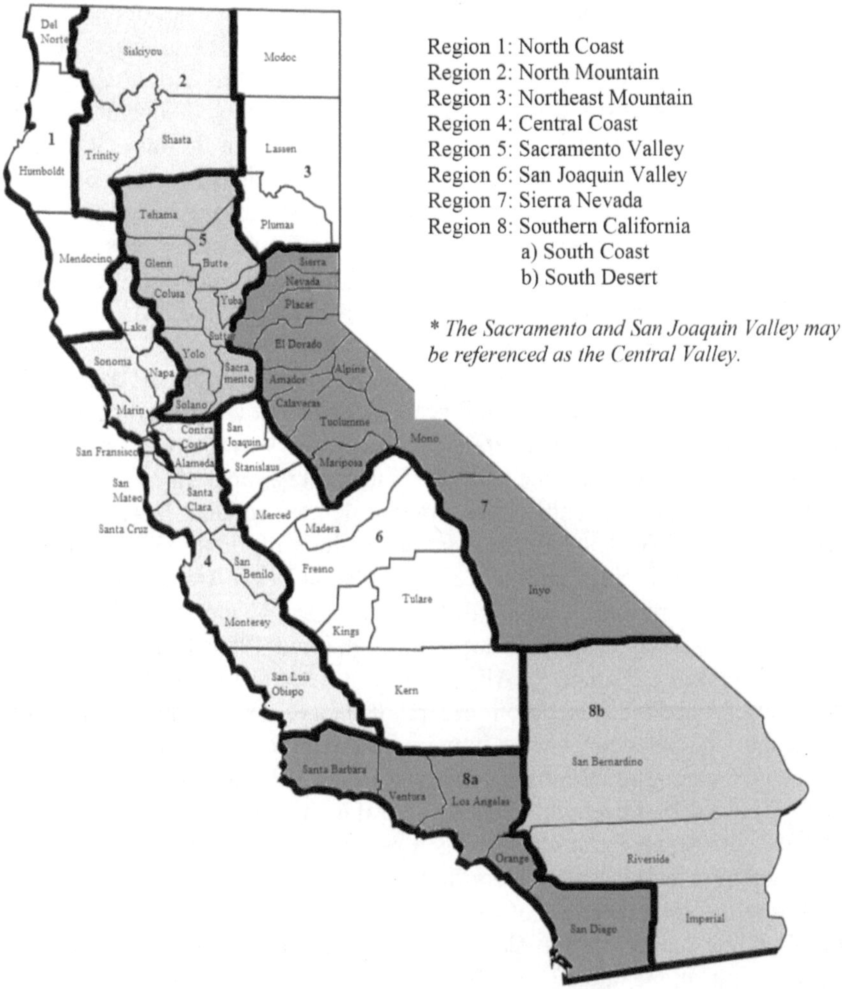

Figure 4.1. California's Eight Agricultural Production Regions
Johnston (2003)

of paid labor. However, not all California farmers were eager to adopt new, innovative farming practices post–World War II or had the ability for capital-intensive farming that is a hallmark of conventional agricultural practices.

California Organic Agriculture

The story of California organic agriculture is just as remarkable and distinctive as the rest of California's agricultural history. According to Guthman,

California contained, "one of the most countercultural branches of the organic farming movement" (Guthman 2004, 14). Organic farming emerged in both rural and urban epicenters predominantly situated in the Central Coast and Central Valley, where the health food movement emerged several decades previously (Guthman 2004, 15). The counter-culture in the late 1960s and 1970s deeply influenced cuisine served in San Francisco while Berkeley became notable because of Chez Panisse, a French-inspired restaurant that opened in 1971 and serves fresh, local, and sustainable ingredients (Belasco 2006; Chez Panisse 2012; Guthman 2004, 15). Two universities, the University California-Santa Cruz and the University of California-Davis, were also significant to the emergence of organic agriculture.

Officially established in 1980, the University of California-Davis Small Farm Center focuses on the small-scale farm operations including foci on specialty crops and rural cooperatives. The center was the result of several years of amorphous existence and efforts to create a program on the campus that served small-scale rather than large-scale conventional farmers (Christensen et al. 1990).[2] The "Tomato Harvester" lawsuit, filed by the California Agrarian Action Project and the California Rural Legal Assistance in Alameda County, spurred the creation of the Small Farm Center at the University of California-Davis campus (Martin and Olmstead 1985; Community Alliance with Family Farmers [CAFF] 2013). Nineteen farm workers claimed, "publicly funded mechanization research displaces farm workers, eliminates small farmers, hurts consumers, impairs the quality of rural life, and impedes collective bargaining" (Martin and Olmstead 1985).[3] While the "Tomato Harvester" lawsuit was eventually lost on appeal in the California Supreme Court in 1989, the Small Farm Center was a response and surviving legacy from legal action taken by stakeholders from the Central Coast.

Before the establishment of the Small Farm Center, Alan Chadwick was the founding leader of "the Garden" on the University of California-Santa Cruz campus in the late 1960s. The Garden was an experimental agricultural endeavor intended to give students a sense of place, "to offset students' feelings of displacement and disruption" (Center for Agroecology and Sustainable Food Systems [CASFS] 2010a). Not only was the Santa Cruz campus being physically redesigned, leaving little outdoor space for reflection, but students were also seeking refuge from political and cultural struggles of the 1960s. Chadwick, a student of biodynamic farming legend Rudolf Steiner, used a French-intensive and biodynamic style farming that would bridge the needs of the mind, body, and spirit. Students and non-students alike were attracted to the Garden with many volunteers later operating their own farms. For example, Orin Martin, who would manage the Farm and Garden[4] starting in the late 1960s, left behind his counter-culture activism activities, like protesting the Vietnam War, in the Northeast. He drove across country to Santa Cruz and eventually started volunteering at the Garden. As he recalls,

> I had some friends who were going to school at UC Santa Cruz. They said, "It's really nice here. Why don't you come out?" This was in 1969. I got a VW bus and drove out here. And there were two people that were living in their house that worked at this place called "The Garden" on campus, and they were never there, I mean, rarely there. They came in late and left before light. So after a while, I thought I would investigate where this garden place was. (Martin 2008, 6)

The Farm and Garden at the University of California-Santa Cruz would eventually become the Center for Agroecology and Sustainable Food Systems (CASFS). Many individuals that would play a key role in the development of California's organic legislation and the development of the organic agriculture market, including Mark Scrowcroft and Mark Lipson, who either attended or were a part of the network surrounding the Santa Cruz campus.

At about the same time the Farm was established in Santa Cruz, the California Certified Organic Farmers (CCOF), a farmer-led chapter-based certification organization, was organized and operating out of Barney Bricmont's dining room (Brians and Brians 2007, 54; Bricmont 2007, 1; California Certified Organic Farmers [CCOF] 2013a). Bricmont was one of six organic farmers from central California intent on creating a certification process for organic farms and products to define and protect the term organic from fraudulent use after efforts by the Rodale Institute, a rural Pennsylvania organic research group, failed to establish a certification program in California. For much of the 1970s and 1980s, CCOF was the only, or at least the predominant, certifier of organic operations in the state of California. In 1992, CCOF leading staff members, including Mark Scrowcroft and Mark Lipson, left CCOF to run the Organic Farming Research Foundation (OFRF). OFRF was created by CCOF in 1990 as a way to provide funding for organic farming research. CCOF leaders did not feel that universities were doing enough to support organic farmers at the time and created OFRF as a vehicle to accept donations for research funding (Lipson 2007, 22–23).

During the lifetime of CCOF and OFRF organizations, certified organic farms and sales dramatically increased. After a watermelon scare in the summer of 1985,[5] CCOF saw membership nearly double (Lipson 2007, 11; Scrowcroft 2007, 47). The demand for membership in CCOF would continue, even from those outside of the state and country, into the late 1980s and 1990s, as more attention was brought to the issue of organic food and agriculture through a number of focusing events, and consumer curiosity and demand (Scrowcroft 2007, 37–40, 42; Lipson 2007, 19; Lipson 2013). Figures 4.2 and 4.3 demonstrate the rapid increase in organic acreage and

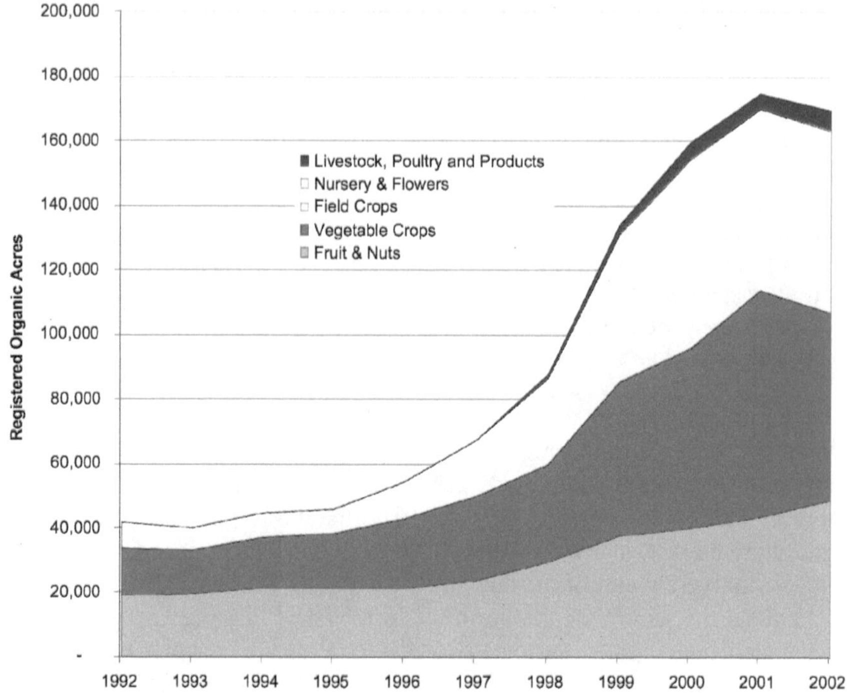

Figure 4.2. 1992–2002 California Organic Acreage
Klonsky 2003, 253 (Reprinted with Permission)

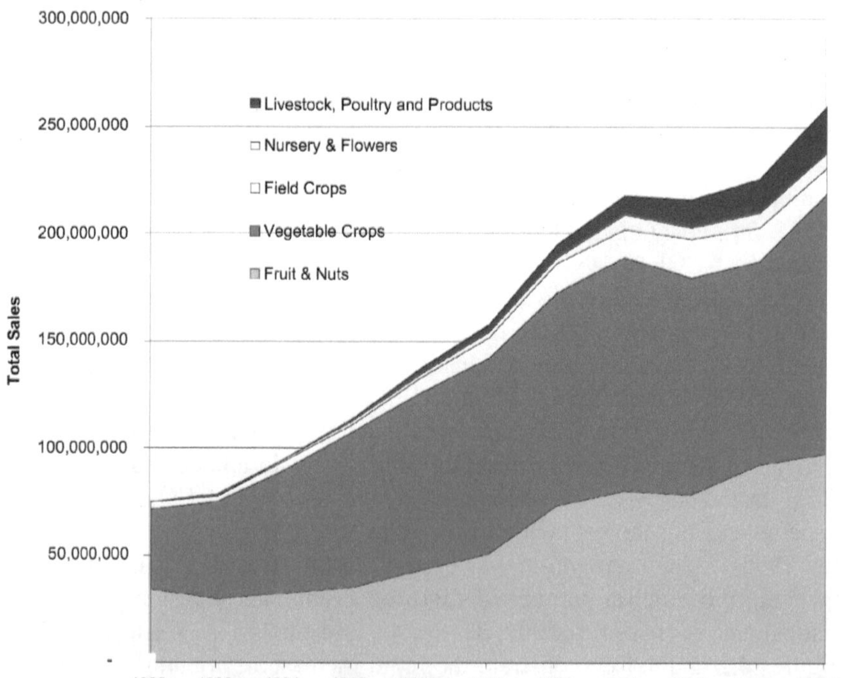

Figure 4.3. 1992–2002 California Organic Sales
Klonsky 2003, 254 (Reprinted with Permission)

sales in California from 1992–2002 with most of the increase attributed to specialty crops such as fruits and vegetables. In 2011, California had 2,530 certified organic operations in the state totaling 951,356 acres of organic crop, pasture, and rangeland (ERS 2013a).

ORGANIC FOOD AND AGRICULTURE REGULATORY LEGISLATION IN CALIFORNIA

California passed a total of eighteen legislative bills regarding organic food and agriculture regulation from 1979–2010. Table 4.1 outlines adopted legislative bills, year of passage, and bill sponsorship. Three legislative bills were the most significant to developing organic food and agriculture policy in the state as each provides a primary framework in which subsequent amendments were made. The Organic Foods Act of 1979 was the seminal piece of legislation while the California Organic Foods Act of 1990 and the California

Table 4.1. California Organic Food and Agriculture Legislation

Year	*Bill*	*Sponsored/Introduced**
1979	AB 443 *Organic Foods Act of 1979*	Michael Gage [D]
1982	AB 3422	Sam Farr [D]
1982	SB 1273	Henry Mello [D]
1990	AB 2012 *California Organic Foods Act of 1990*	Sam Farr [D]
1991	AB 645	Sam Farr [D]
1992	AB 3246	Sam Farr [D]
1993	AB 1713	Sam Farr [D]; Rusty Areias [D]
1994	AB 2518	Rusty Areais [D]
1995	SB 688	Committee on Agriculture and Water Resources
1996	AB 2340	Sal Cannella [D]
1998	AB 2761	Committee on Agriculture
1999	AB 1243	Committee on Agriculture
2002	AB 2823 *California Organic Products Act of 2003*	Virginia Strom-Martin [D]
2003	AB 776	Barbara S. Matthews [D]
2005	SB 730	Jackie Speier [D]
2009	AB 856	Anna Caballero [D]
2010	AB 2612	Committee on Agriculture
2010	AB 2686	Tom Berryhill [R]

**Political Party – [D] = Democrat; [R] = Republican*

Source: Data compiled by author.

Organic Products Act of 2003 each represent a significant overhaul and reform. The other thirteen bills represent either reauthorization, as in the case of AB 3422 in 1982, or were intended for refinement and clarification.

From 1979–1989, regulations were only developed within the Health and Safety code. After 1990, organic food and agricultural regulations existed in both the Health and Safety code and the Food and Agricultural code spanning implementation responsibilities across two different state agencies. The California Department of Health Services (CDHS) is responsible for enforcement activities for the Health and Safety code. The California Department of Food and Agriculture (CDFA) is responsible for enforcement activities for the Food and Agricultural code. While this study does not explicitly examine the development of state administrative agency rules, the role of CDHS and CDFA are significant to the development of legislation and overall California organic policy design.

The history of organic food and agriculture legislation in California is explored in three different eras. First, the passage of the Organic Foods Act of 1979 and reauthorization in 1982 is discussed. Then the events leading up to and the passage of the California Organic Foods Act of 1990 are described. Amending legislation from 1991–1999 will also be covered, as well. Finally, the history will conclude with the passage and amendments of the California Organic Products Act (COPA) of 2003. It is during this latter period that California's law challenges federal policy standards.

The Organic Foods Act of 1979 (AB 443)

The organic food and agriculture legislation passed in 1979 was not the first attempt by the California State Legislature to pass a law of that nature. In 1978, Assemblyman Vic Fazio, a Democrat representing constituents in the central California region, authored AB 2135, a bill that "proposed definitions and controls" for the term organic in California agriculture (Legislative Counsel 1979). The bill was proposed at the recommendation of the Department of Consumer Affairs under the pressure of Governor Jerry Brown, a Democrat. While the bill died after the third reading on the senate floor, Assemblyman Michael Gage, a Democrat representing Napa, Sonoma, and Vallejo, would sponsor AB 443 in 1979. AB 443 was very similar to AB 2135. AB 443's intent was to establish "standards within the Sherman Food, Drug and Cosmetic Law for food products being advertised or labeled as organic, organically grown or wild, natural grown, ecologically grown, or biologically grown" (Legislative Counsel 1979). Essentially, the bill was intended to provide a definition of organic to eliminate confusion and deceptive practices in the market. Eliminating an information asymmetry would provide consumers a clear indication for the sort of product they would purchase.

The final version of AB 443 altered the Health and Safety code to include sections 26469, 26569, and 26850 relating to organic food. Within these sections, key definitions are made, acceptable practices are identified, implementation funding determined, and a sunset clause established. See table 4.2 for key aspects of the bill. The bill offers precise guidelines for permissible use of terms and conditions of growth, production, and handling of organic

Table 4.2. Summary of AB 443/Cal. Hlth & S. Code § 26469, 26569, 26850

Aspect	*Summary*
Definitions and Use	Identifies the appropriate use of the terms organic, organically grown, naturally grown, wild, ecologically grown, and biologically grown. Limits use to "raw agricultural commodities or processed foods, which have been produced without the use of synthetically compounded fertilizers, pesticides, or growth regulators for twelve months before seed was planted until the time of harvest, or harvested, stored, processed or packed" (Fishman 1979a). Processed foods can contain ascorbic acid, sodium ascorbate, calcium ascorbate, and citric acid. Synthetically compounded means "products formulated by a process which chemically changes a material of substances" (e.g., splitting molecules) from a natural source except microbiological processes that only physically change the structure (Cal. Hlth & S. Code § 26569.11 [a][3]).
Pesticide Residue	The maximum level of acceptable pesticide residue for organically grown food is 10 percent of the level determined safe by the United States Food and Drug Administration (Cal. Hlth & S. Code § 26569.12).
Meat, Fish, & Poultry	Meat, fish, and poultry sold as organic can be treated with chemicals or drugs for disease. However, chemicals and drugs cannot be used within ninety days of slaughter, and 60 percent of the final sale weight must result from organic feed (Cal. Hlth & S. Code §26569.11 [d]).
Acceptable/ Unacceptable Materials	Permissible materials in organic production include calcium oxide, sulfur, gypsum, light petroleum, and vegetable oils. Unacceptable materials include Bordeaux mix, aromatic petroleum solvents, diesel, petroleum fractions used as weed and carrot oils (Cal. Hlth & S. Code 26569.11 [a][3]).
Record-Keeping	Growers, manufacturers, wholesalers, and retailers much keep accurate records for two years regarding source and details of food products labeled organic (Cal. Hlth & S. Code §26569.11, 26569.15 [c-e]).
Certification Labeling	Food products that are certified organic and not simply labeled organic must list the person or organization that provided certification on the label (Cal. Hlth & S. Code §26569.15)
Enforcement	No funds are allotted for enforcement activities. No state agency is required to enforce the code (Cal. Hlth & S. Code §26850.5 [c]).
Sunset Clause	Each section was set to expire, unless reauthorized, on January 31, 1983 (Cal. Hlth & S. Code §26469, 26569, 26850).

Source: Data compiled by author from state bill and statute information.

food. However, certification of organic operations would remain voluntary. CCOF members wanted to keep certification as farmer-to-farmer certification as the method for validating organic claims and agreed to keep terms of agreement voluntary (Cantisano 2008, 113). In addition to voluntary compliance, enforcement activities would not be mandatory for state agencies including CDHS. Citing the adoption of Proposition 13, it was expected that there would be cutbacks in government funding (Fishman 1979a).[6] Therefore, mandating state agencies to enforce the law without proper additional funding and resources was thought as inconceivable.

According to Bones (1992), Oregon not only provided the momentum for organic food and agriculture legislation in other states but also used Oregon's 1973 administrative rules as a model. The California law differed from Oregon rules in record-keeping requirements, defining synthetic inputs, and labeling requirements. At the time, California's rules were not only similar to rules developed in Oregon and Maine, moreover, California's definition of organic was in line with definitions developed by the Federal Trade Commission in 1979 (USDA Study Team 1980, 6). Regardless of similarities, California's adopted legislation in 1979 sheds light on the complexities for building consensus around initial efforts to regulate the market. The link between Oregon and California was established through an organic network including CCOF, Oregon Tilth, and Friends of the Earth among other self-organized groups nationally like the Northeast Organic Farming Association. These groups worked to build consensus and minimize opposition to legally defining and establishing organic rules.

Support and Opposition (AB 443)

There was little evidence to support any political disagreement within the California State Legislature. Instead, most of the disagreement occurred between food and health industry groups. AB 443 divided organic farmers and interested consumers into two camps. One camp feared the law as being too strict and disliked government intervention in this area. The other camp was also cautious of government involvement but was determined to fight against fraud and the watering-down of the term "organic" (USDA Study Team 1980, 8). CCOF was in the latter camp and played an instrumental role in developing and advocating AB 443. Sy Weisman and Stuart Fishman,[7] both with the North Coast chapter of CCOF and Gage's constituents, initially expressed reservations about AB 2135 in its treatment and regulation of the term organic. The primary concern was "it would exclude virtually all of the growers using the term" at the time (CCOF 1988, 2).

In rewriting AB 443, CCOF members worked extensively to develop a definition of organic and set of regulations that the California organic industry could agree upon. Fraud and loose application of the term organic on agriculture produce was a major concern. Working at focusing events, an article in *New West* magazine and a Los Angeles CBS television news affiliate each exposed fraud in the organic market in 1979 (Fishman 1979b). Conventional growers could have an unfair market advantage by selling conventionally grown produce as organic. This sort of fraud damaged the image and integrity of the organic market. Many in the organic industry linked organic fraud with the diluting of the term natural. The term natural was perceived to have lost all meaning, and the effort was made to prevent the same diluting of the term organic through revising AB 443 (Fishman 1979b). AB 443 represented a method for identifying what farming operations were or were not organic.

The distinction on who was organic and not organic was a process that took two years. CCOF started working on AB 443 in 1978 (CCOF 1988). Meetings were held with organic growers, processors, wholesalers, and retailers to develop a law that would codify the organic term. Support was drawn from a wide base including the California District Attorneys Association, Department of Consumer Affairs, California Health Officers Association, Sacramento Natural Foods Co-Op, and California Dietetic Association (Legislative Counsel 1979). From outside of California, CCOF worked closely with Friends of the Earth and Oregon Tilth (Fishman 1979a). The result ended in an eight-page bill detailing the terms and agreement of using the word organic. The final bill version was a clear product of compromise and negotiation with opposing forces. While still a victory, CCOF was not completely satisfied with the passed version (CCOF 1988).

Opposition to the bill came from two primary sources[8] outside of the organic industry network—Clinton Miller representing the National Health Federation and Dave Ajay from the National Nutritional Foods Association. Miller called supporters of the bill "naïve but well meaning" (Miller 1978); Ajay had several key points of criticism (Lasher, n.d.[a]). First, voiced opposition demonstrated a clear dislike for the organic definition included in the bill. Miller claimed the organic definition used in the bill was only one of many definitions used by organic producers. Moreover, the claim was made that the organic definition was arbitrarily different from the definition of natural (Fishman 1979c; Lasher, n.d.[a], n.d.[b]; Miller 1978). Ajay contended that having multiple state definitions for organic would be problematic and suggested waiting for the Federal Trade Commission to develop regulations, which would actually never come to fruition.

A second concern was centered on quality differences between organic and conventional agriculture. Miller believed the bill should contain language to suggest organic and natural foods were superior in quality to conventionally grown foods. Quality refers to higher nutritional value as well as beneficial to health and the environment because of the quality of soil and absence of mechanized inputs (Fishman 1979c; Lasher, n.d.[a], n.d.[b]). Ajay elaborated on the quality point by arguing that the twelve-month cessation period to transition from conventional to organic production was not enough to protect soil and product quality (Fishman 1979c).

The third and final key criticism centers on the involvement of CDHS. Miller claimed that involving the department would be bad for natural food stores, growers, and consumers. In a conversation with Ellen Powell, the legislative coordinator for the Department of Consumer Affairs, Miller indicated that the, "hostility 'tween the health food industry and the DoH goes back a long ways, and underlies A[jay] and M[iller]'s worry of the bill" (Lasher, n.d.[a]). The perceived hostility was because CDFA implemented the Sherman Food, Drug, and Cosmetic Act. Miller's assumption was not clearly linked to Proposition 13 concerns or CDHS's reservations about funding (California Department of Health Services [CDHS] 1979).

In addition to Miller and Ajay's voiced opposition, some organic farmers also expressed concerns. Indeed, Bill Jessup, a citrus grower from Oasis, indicated reservations about the need for government intervention. In a letter written to Stuart Fishman, Jessup states, "I haven't gotten a look yet at AB 443. My first inclination is that it would be better to rely on private regulation than on the government" (Jessup, n.d.). Jessup sought a more stringent definition of organic and wanted to prevent having multiple definitions across the country. He felt that multiple definitions would cause confusion and could even be more damaging than the deception and fraud the industry was currently fighting.

The voices of opposition highlight differences between the consumer and producer perspective of organic regulations highlighted (Clark 2015; Obach 2015). From a consumer perspective, government involvement would likely lead to a singular and narrowly defined definition for organic. It would also be possible for conventional and industrial interests to have a greater influence in standards development. The producer perspective, to the contrary, eliminated some of the potential for industrial and conventional influence by maintaining a wider range or scope of permissible organic practices. Instead, a singular definition or approach to producing organic foods, a farmer could be more or less organic. The producer perspective is not as consumer friendly, as it lacks a singular and specified approach, but the perspective does permit varying approaches to organic production that could accommodate different farming scales, philosophies, and production preferences.

Compromise (AB 443)

To address some of the voiced opposition, the compromises to AB 443 included a sunset provision, removal of a definition for natural, technical exceptions to using the term wild, and the inclusion of language regarding no expectations of enforcement. The enforcement language (Cal. Hlth & S. Code §26850.5[c]), or rather the language indicating no mandatory enforcement, was included to address concerns raised by CDHS regarding funding support. In an earlier bill analysis, CDHS indicated a neutral position because of a lack of funding (CDHS 1979).

The removal and exceptions related to term definitions for natural and wild were made in light of opposition. As described in Section 26469, the term wild can be "used to describe a flavor of a food or a plant variety." This language was added after concerns related to naming of wild cherry cough drops raised by the California Association of Tobacco and Candy Distributors. There was no intent to change this long-time tradition (Fishman 1979c). The exclusion of the term natural from being regulated (see Cal. Hlth & S. Code §26469) is the result of a trade-off with agribusiness (Fishman 1979d). By not regulating natural-labeled foods, industrial and conventional interests agreed to leave the organic label alone.

Perhaps one of the bigger compromises involved the establishment of a sunset provision for January 31, 1983. Designers of the bill wanted to set up a regulatory board under the law that would add and modify permissible materials and practices (Fishman 1979d). It was understood the law would contain loopholes that would need to be addressed at a later date. However, the regulatory board was not approved. Therefore, the sunset provision was "unanimously and enthusiastically agreed upon" and included in the final bill version. It would allow concerned parties to create a more workable bill after discovering any loopholes or deficiencies (Fishman 1979a).

1982 Reauthorization (AB 3422)

AB 3422, introduced by Assemblyman Sam Farr, was the reauthorization and amending bill to the Organic Foods Act of 1979. AB 3422 was debated after the passage of SB 1273, an urgency measure to create exceptions to the maximum allowable pesticide residue on organics after a medfly infestation.[9] After the two-year trial period for AB 443, several deficiencies and areas of improvement were identified. There was enough support for the continual existence of organic regulations despite continual in-fighting within in the organic community (Cantisano 1982). Some recommended improvements ranged from amending the list of permissible materials, required documentation for soil improvement, funding CDHS to improve enforcement, and giving distinctive definitions for ecologically or biologically grown foods (Castisano 1982). After a lengthy discussion for how to improve the existing law, AB 3422 eliminated the sunset provision and outlined new limitations to applying prohibited substances to raw agricultural commodities. Efforts to increase the "teeth" or enforcement of the bill were unfulfilled. The CDFA was hostile to the idea of getting involved with organic regulation, and the CDHS did not want to be responsible for enforcement without a funding source (CDHS 1982).

The California Organic Foods Act of 1990 (AB 2012)

In March 1989, Sam Farr introduced AB 2012 or The California Organic Foods Act for the first time (Lipson 1990). The passage of the act involved two years of effort to craft a bill that would overhaul organic regulations. The 1980s marked a decade of both disappointment and joy in the organic community. Although the interest in organic products was increasing, weak enforcement of the 1979 law led to numerous scandals and the need to create mandatory labeling standards. CCOF, one of the primary sponsors and advocates of the bill, aimed, "to increase the regulation of the labeling of organic foods and the inspection of foods grown under conditions considered as organic to meet the public's expectation and prevent fraud" (Evans 1989).

The Call for Reform in California

The movement for reform in California was spurred in the 1980s as attention to organic food and agriculture was increasing. A few high-profile cases of fraud in the market increased concern about the image and validity of the organically labeled goods. Meanwhile, an increasing tide of concern over conventional agricultural practices led to increased consumer demand and curiosity. The carrot caper scandal was one of the biggest fraud scandals that highlighted the issue of marketplace fraud (Lipson 2007, 19; Scrowcroft 2007, 42; Epstein 2013). In 1988, photographic evidence surfaced incriminating Pacific Organics, a food distributor, of re-bagging conventionally grown Mexican carrots as organic (Scrowcroft 2007, 37–38; CCOF 2013a). The "organic" carrots were then distributed to a number of grocery outlets and sold under false pretenses to organic consumers. CCOF and CDHS were investigating the odd occurrence of organic carrots being sold in Weed Patch, California, when the primary organic carrot grower was out of carrots. After the help from a Pacific Organics worker, evidence showed trucks from Mexico bringing conventionally grown carrots entering a warehouse and being re-bagged as organic carrots. According to Scrowcroft (2007, 38–39), CCOF went public with information in the *San Jose Mercury News* with the Associated Press picking up the news story shortly thereafter. Cases of fraud like the carrot caper scandal were not uncommon. However, the carrot caper incident was one of the more blatant cases leading to considerable concern in the organic community about protecting itself.

Another wave of concern emerged from public and environmental health issues regarding food contamination and questionable conventional agricultural practices. One focal point of attention came during the February 1989 broadcast on CBS's *60 Minutes* examining the dangers of Alar and other pesticides (Negin 1996; Scrowcroft 2007, 35). The broadcast was based on the Natural Resources Defense Council's report "Intolerable Risk: Pesticides in Our Children's Food" (Negin 1996). In response to the broadcast and the involvement of American actress Meryl Streep, the public panicked and the number of organic farmers grew drastically (Negin 1996; Scrowcroft 2007, 35). Alar, also known as Daminozide, is a chemical sprayed on apples that enhances color and regulates growth. It is a known carcinogen being banned in California under Proposition 65 on January 1, 1990 (Office of Environmental Health Hazard Assessment 2013). The focus on pesticides, not just Alar, at the time coincided with the passage of Proposition 65, a California initiative for labeling hazardous chemicals in the marketplace (Lipson 2013). Heightened concern about the dangers of conventional agriculture led to increased interest and demand of organic food and agriculture.[10]

Calls for reforming organic food laws in California came from the convergence of concerns over fraud and health concerns. As Bricmont recalls about the need for reform,

> Originally we went on our own code of ethics. But we had no way to control the industry. Anybody could just slap a label "organic," because organic meant nothing, there was no legal definition. At that point we were running across people who were just chemically growing, and labeling the food organic and selling it. We knew we had to clean up the industry. We knew what we were doing as a group. But then the outside world was not cooperating. So we decided we needed to draw a line in the sand that said: on this side of the line you're organic, on the other side you're not. (Bricmont 2007, 8–9)

CCOF, who was truly the de facto enforcer of the 1979 law, led the effort for reform by hiring attorney Barry Epstein and working with Assemblyman Sam Farr to rewrite the law (Brians and Brians 2007, 51; Lipson 2013; Epstein 2013). Epstein was an environmental attorney in San Francisco that was introduced to the organic farming community during the 1988 medfly eradication, when he represented a group of organic farmers in Northern California who were fighting against the mandatory medfly eradication spray program in 1988. His expertise was sought for re-writing the bill partially at the request of Farr. Farr was approached by what he describes as "hippie farmers" from his district seeking a legal course for enforcement (Farr 2007 12–13; Farr 2013). He worked to educate the farmers about the legislative process and how to go about writing and getting a bill passed. In addition to contacting key decision makers in the legislative process, part of that effort was to work with lawyers to write the bill in codified language (Farr 2013). Epstein worked closely with Mark Lipson and the CCOF grower committee to write the law starting in 1989 (Lipson 2013). After two years of meetings[11] and a number of revisions, the bill was reintroduced on March 9, 1990 after action during the previous legislative session failed.[12]

Summary of the Final Bill Version (AB 2012)

The passed version of AB 2012 was approximately 13,500 words and was unusually detailed. The bill added chapter 10, Section 46000 and Section 14904 to the Food and Agricultural code and altered the Health and Safety code by amending and moving regulatory code to Article 4.5 Section 26569.20. Table 4.3 outlines major provisions of the bill. Adoption of the bill extended mandatory enforcement requirements for CDHS and CDFA. Fees would be collected from registration of persons engaged in the production or handling of organic goods and used for enforcement activities. While application limits and some term definitions were provided, AB 2012 required the establishment of the Organic Food Advisory Board. The board would be responsible for advising the CDFA director and developing regulations regarding permissible and impermissible substances and practices. Prior to regulation adoption, the CDFA director is responsible for publishing a list of materials, permissible and impermissible, based on materials published by CCOF, the Organic Food Production Association of impermissible, based on materials published by CCOF, the Organic Food Production Association of North America, and Oregon and Washington Departments of Agriculture (Cal. Food & Ag. Code §46003.5).

Certifying organic operations, a voluntary measure, would not be the responsibility of CDFA or CDHS. Instead, a third-party certification approach was maintained from the 1979 bill. However, certifying agencies would be responsible for registering with the state and maintaining records. Furthermore, the certifying agency would need to be legally separate and independent from any entity it certified.

Support, Opposition, and Compromise (AB 2012)

The debate and compromise surrounding the two-year development of AB 2012 centered on three different groups of participants including elected officials, public organizations, and the agricultural industry, including both

Table 4.3. Summary of AB 2012

Aspect	*Summary*
Enforcement	The bill requires that the Director of CDHS and CDFA to enforce regulations regarding organic foods (Cal. Food & Ag. Code § 14904; Cal. Hlth & S. Code § 26569.25).
Registration & Fees	All persons engaged in the production or handling of organic production and handling ARE required to register with the agricultural commissioner annually. For producers, registration fees are on a sliding scale based gross income (Cal. Food & Ag. Code § 46002; Cal. Hlth & S. Code § 26569.35).
Application Limitations	The bill identified the permissible time frames in which pesticides and other prohibited materials could be applied to crops, seeds, and growing mediums if identified and labeled as organic. A one-year transition period was established (Cal. Hlth & S. Code § 26569.22).
Term Definitions	The bill included precise definitions for such terms as applied, handled, processed, prohibited materials, sold as organic, and synthetically compounded (Cal. Hlth & S. Code § 26569.21–.22).
Advisory Board	The CDFA Director is responsible for the establishment of the Organic Food Advisory Board that will advise the Director and adopt regulations regarding permissible and prohibited materials. The board is a 13 member body representing producers, processors, retailers, consumers, scientists, and environmentalists (Cal. Food & Ag. Code §46003).
Certification	For products sold as certified organic, products must be certified by a registered organization that is a legally separate and independent organization than the entity being certified (Cal. Food & Ag. Code § 46009; Cal. Hlth & S. Code § 26569.30 & §265569.24).

Source: Data compiled by author from state bill and statute information.

conventional and organic agriculture. Conventional agricultural interests, including the California Farm Bureau and the Western Growers Association, were slow to recognize the significance of the bill, and, after their strategy of trying to kill the bill failed, tried to make the bill workable (Farr 2013; Epstein 2013). Three key points of debate and compromise center on the complexity of the bill including third-party certification, funding, and technical aspects of regulation.

One of the more notable aspects of the bill is its sheer length and complexity. At approximately 13,500 words, the bill is comparatively longer than most legislation and remarkably more like administrative rules. As noted in a California Department of Finance bill review, AB 2012 establishes "a program that is complex, with program responsibilities spread between both State and local agencies including the (C)DFA, the (C)DHS, the CACs (county agricultural commissioners), and local district attorneys" (California Department of Finance 1990). The detailed complexity of the bill, however, was a calculated move by CCOF and its supporters. According to Barry Epstein (2013), the reason for the extensive detail was because of CCOF's concern about the rule-making process with CDHS and, more specifically, CDFA. The perception was that CDFA was captured by conventional agricultural interest. Had the bill passed delegating authority to CDFA and CDHS, rules may have been developed that did represent the best interests of CCOF or the rest of the organic agriculture community.

The third-party certification arrangement was one of the more hotly contested aspects of the complex bill. One reason the third-party certification scheme was included was related to politics. As Sam Farr pointed out (2007, 16–17; 2013), Republican governor George Deukmejian posed as a challenge for passage.[13] The third-party certification design worked only if it did not increase the cost or regulatory burden at a time of anti-regulation in the United States. Yet, another challenge arose from state and local government entities wanting to retain regulatory power by eliminating the third-party approach. CDFA, county agricultural commissioners (CACs), and CDHS, all at one point during the process raised concern about the third-party scheme. From an agricultural standpoint, it was assumed that CACs could conduct independent certification (Kizer 1989; Voss and Nutter 1989; Epstein 2013). Likewise, CDHS, who would support the bill in later versions, indicated that the AB 2012 "would place the DHS in direct competition with the private industry in the certification of organic food processors, handlers and producers" (CDHS 1989; Kizer 1989; Voss and Nutter 1989). Giving support to state agencies' position were a number of interest groups such as the California Cattleman's Association (1989), which cited private certification as a barrier for entry into the market. Both state agencies initially failed to recognize the importance of the farmer-on-farmer certification process for securing passage.

As a second and more well established reason for the third-party certification was the demand from CCOF. CCOF growers truly believed in third-party certification but did not feel it should be mandated in the new law despite fraud. As Mark Lipson recalls:

> The board of CCOF didn't feel like it was right to impose a requirement for certification. They wanted that to be their additional level of value in the marketplace. But they said, "There needs to be some kind of legal baseline, so we'll create this registration program where everybody has to basically declare that they're following the law and make a legal affidavit that they're following the standards in the law. But then, third-party certification will be over and above that." (Lipson 2007, 20)

AB 2012 wouldn't diverge too much from the Organic Foods Act of 1979 except that enforcement and registration would be mandatory. Certification from a third-party entity would remain as the method for label certification but with more clear indication for what constitutes as organic. The third-party certification scheme not only met the desires of CCOF, but also the cost-effectiveness requirements and infrastructure requirements set forth by Governor Deukmejian (Lipson 1990).

Another significant barrier to the passage of AB 2012 was ensuring appropriate levels of funding for CDFA, CDHS, and CACs.[14] On one side of the debate were a number of organizations against registration fees. For example, the California Grocers Association (1989) was against fees charged to retailers selling organic food. As stated previously, Governor Deukmejian posed a challenge for passage in part due to cost requirements of implementation (Farr 2013). The California Department of Finance, CACs, CDHS, and CDFA also indicated concerns, particularly early on in the process in 1989, about the cost of implementation and enforcement (CDHS 1989; California Department of Finance 1990; Lipson 1990). There was uncertainty regarding initial start-up cost and long-term viability of the program to sustain itself. If AB 2012 were to pass, it would have to contain funding to cover enforcing the bill.

Registration fees collected from organic producers and handlers were the primary vehicle to fund enforcement (Lipson 1990).[15] However, additional compromises were made to ensure adequate funding levels. Specifically, writers of the bill "trimmed certain functions called for in the bill, phased in the start-up of some programs, and limited the scope of rule-making activity that is required" (Lipson 1990). For example, failure for a certifying organization to register would be a violation after December 31, 1991, allowing a one-year deferment of enforcement. While not all parties found the scenario ideal, the effort to "balance the equation" helped to overcome the financial problems associated with the passage.

Controversy also surrounded technical aspects of AB 2012, ranging from improper coding, record keeping requirements, and decisions regarding organic practices. Revisions were made to protect the identity of registrants, require growing practices be disclosed during registration, permit the use of the 1990 CCOF Certification Handbook as a reference for prohibiting specific substances until 1992, and reduce the tolerance level from 10 percent to 5 percent for prohibited material contamination on organic products (Lipson 1990). These technical adjustments were just some of the many technicality compromises made.

At the time technical aspects of AB 2012 were developed, an effort was forged to standardize organic practices. The Western Alliance of Certification Organizations (WACO), including representatives from Oregon Tilth, Washington Tilth, and CCOF, was created and active since the late 1980s (Lipson

2007, 18; Lipson 2013). WACO's purpose was to "harmonize" organic standards and practices on the west coast. In 1990, California was operating on a one-year transition period to organics while Oregon and Washington were operating on a three-year transition period (Scrowcroft 2007, 40). The difference between transition periods caused Oregon to consider refusing California's organic products. Such discrepancies led to a handshake agreement to iron-out differences. Over the course of several meetings, which also had participants from the Northeast Organic Farming Association chapters, an agreement was made to standardize regulations to improve interstate commerce and minimize competitive disadvantage nationally (Epstein 1989b; Scrowcroft 2007, 40–41). While not all technical changes in AB 2012 synchronized with other practices and standards, several legislative adoptions in the 1990s would work toward standardizing practices and norms among states.

Amending Adoptions of the 1990s

A total of eight bills were passed from 1991–1999, amending provisions of the California Organic Foods Act of 1990. Table 4.4 provides a brief description of each bill's intent. Of all the bills, the 1992 AB 3246 was the only bill to be vetoed prior to passage. Republican governor Pete Wilson cited inadequate funding for CDHS to implement the law (Legislative Counsel 1992). No other bill was found as vetoed during its course to passage.

Three patterns emerged for amending legislation that occurred during the 1990s aside from implementation responsibility adjustments. First, five bills contain language that directly impact the scope and size of the Organic Food Advisory Board. The 1991, 1992, 1993, and 1996 bills extended the deadline in which the board should adopt regulations regarding prohibited materials. The 1993 bill extended the final date to January 1, 1996. All of the amendments were intended to aid in the continual implementation of the California Organic Foods Act of 1990, but the 1993 extension was explicitly noted in response to delayed federal efforts to the development of the prohibited substance list at the federal level. As stated in a hearing,

> It is the intent of the Department of Food and Agriculture to adopt organic food regulations in a form and content similar to those adopted by the U.S.D.A. pursuant to the federal Organic Foods Production Act of 1990. Inasmuch as those regulations will not be prepared until 1994 or 1995, the extension of the time limit for the state's adoption from 1994 to 1996 appears reasonable. (Hite 1993)

No further extension of the deadline was made after 1993, once members to the NOSB were appointed, but the 1996 bill changed the deadline to a future date when the national standards are developed.

Table 4.4. Bills Amending the California Organic Foods Act of 1990

Year	*Bill*	*Summary of Intent Relating to Organic Provisions*
1991	AB 645	An urgency measure intended to continuously appropriate revenues, extend phasing deadlines for new regulations and revise established fee schedule for registration.
1992	AB 3246	The bill would alter the complaint procedure, registration fees, and required adoption of prohibited substances be completed no later than July 1, 1994.
1993	AB 1713	The bill amended a number of provisions including the following: add one additional member to the Organic Food Advisory Board; extend the date for listing prohibited substances to January 1, 1996; prohibit prescribe records from becoming public; ban co-mingling of organic and non-organic commodities; protect trade secrets, and limit the multi-ingredient labeling on certain organic products.
1994	AB 2518	The bill amended a number of provisions including the following: allow for approval of certifiers based on national law; increase responsibilities of CACs; require all organic be certified; and corrected a technical glitch regarding application of pesticides in the 1990 bill.
1995	SB 688	The bill clarifies deadlines and requirements for when prohibited materials can last be applied. These changes remain in effect until the implementation of federal organic certification.
1996	AB 2340	The bill changed the deadline for when the prohibited substances list should be adopted.
1998	AB 2761	The bill creates alternatives for the Organic Food Advisory Board.
1999	AB 1243	The bill would further amend membership aspects of the Organic Food Advisory Board, make technical changes to approved ingredients for organic food, and alter the scope of power of the CDHS in implementing the law.

Source: Data compiled by author.

A second pattern within the 1990s amending legislation relates to the membership base of the organic advisory board. In 1993, the board was expanded to fourteen members with the addition of another producer representative. In 1998 and 1999, policy makers created and made technical adjustments for alternative board members. None of the adjustments received significant criticism but led to altering representation dynamics from various industries and constituents vested in the development of California's organic regulatory standards.

A third trend of the 1990s amending legislation regards the moratorium periods of transitioning to organic. In the 1990 legislation, a twelve-month period of transition was required where land may not have prohibited materials applied. The 1994 and 1995 bills clarified a glitch in the original bill and established organic transitionary periods based on initial registration dates. Table 4.5 identifies the transitionary period length based on the registration date. Starting

Table 4.5. Organic Transitionary Periods Established after Passage of 1995 SB 688

Registration Date	*No Prohibited Material Period*
Prior to January 1, 1995	Previous 12 Months
During 1995 calendar year	Previous 24 Months
After January 1, 1996	Previous 36 Months

Source: Legislative Counsel (1995)

in 1996, the three-year transitionary period agreed upon by WACO was achieved in California. These changes received no significant opposition.

The California Organic Products Act of 2003 (AB 2823)

AB 2823, also known as the California Organic Products Act of 2003, was chaptered into law by the California secretary of state on September 15, 2002. Introduced early during the 2002 legislative session by Assemblywoman Virginia Strom-Martin, a Democrat from Duncan Mills in Sonoma county, AB 2823 brought the state organic food program in compliance with the National Organic Program (NOP) effective in 2002. Starting in August 1999, meetings were held by the California Organic Food Advisory Board with interested parties[16] to collaborate on the re-writing of California's organic law with the expectation of bill passage in 2002.[17]

Summary of the Final Bill Version (AB 2823)

The passed version of AB 2823 was as detailed as the California Organic Foods Act of 1990. The bill modified the Food and Agricultural code and the Health and Safety code regarding organic foods. Table 4.6 outlines major provisions of the bill. The bill maintained a mandatory third-party certification system and left enforcement responsibilities to the CDFA, CDHS, and CACs. The Organic Food Advisory Board was replaced with the California Organic Products Committee. The committee would be responsible for advising the California secretary of agriculture regarding his or her duties as assigned by AB 2823. Enforcement and certification responsibilities by all California entities, however, must be in compliance and meet the standards set forth by the NOP.

The California Organic Products Act of 2003 made substantial changes regarding fee structures and scope of materials and products covered by organic regulation. To offset increasing costs, a fee structure was created for CDHS, new fees were created for particular products and handlers, and CDFA fees were increased. The fee increases were at the request of the organic industry to cover certification costs (Strom-Martin 2002a). For the scope of

Table 4.6. Summary of the California Organic Products Act of 2003 (AB 2823)

Aspect	*Summary*
Enforcement & Certification	Requires the CDFA, CDHS, and CACs to enforce regulations (Cal. Food & Ag. Code § 46000; Cal. Hlth & S. Code § 110812). Maintains the third-party certification system as also established by the NOP.
Restructures Code	Brings the Food and Agricultural Code and Health and Safety Code into compliance to the NOP.
Advisory Board	Establishes the California Organic Products Committee to advise the Agricultural Secretary. Comprised of 15 members and 15 alternatives representing producers, processors, wholesale distributor, consumer representatives, environmentalists, scientists, and retailers (Cal. Food & Ag. Code § 46003).
Definitions & Coverage	Maintains most definitions and includes new definitions for animal food and non-food plants, and USDA NOP (Cal. Food & Ag. Code § 46004.1; Cal. Hlth & S. Code § 110815). The bill also extends coverage of organic items to include multi-ingredient items, cosmetics, alcohol, dietary supplements and pet food (Cal. Hlth & S. Code § 110835). Cosmestic and multi-ingredient products must contain at least 70 percent organically produced ingredients (Cal. Hlth & S. Code § 110838-110839).
Registration Requirements	Maintains previous registration requirement and penalties (Cal. Food & Ag. Code § 46013.1; Cal. Hlth & S. Code § 110875 and § 110915).
Fees	Amends CDFA fee structure to increase fees and fee caps and creates a fee schedule for CDHS. Also creates new fees for cosmetic products, multi-ingredient products, and for individuals that otherwise handle organic products not covered by CDHS scope (Cal. Food & Ag. Code § 46013.1(f); Cal. Hlth & S. Code § 110875(d)).

Source: Data compiled by author from state bill and statute information.

coverage, alcohol, cosmetic products, dietary supplements, and pet food would be included as regulated items. Cosmetic products, as well as multi-ingredient products, would be subjected to additional fees. The inclusion of newer items beyond raw agricultural products or processed goods is reflective of the market that was expanding and diversifying beyond just food.

Support, Opposition, and Compromise (AB 2823)

The controversy and debate surrounding AB 2823 departed from previous debates on organic regulations. Except for the 1992 AB 3246, partisanship did not clearly appear to affect the outcome of any organic bill prior to 2002. Even with the passage of the California Organic Foods Act of 1990, Republican governor George Deukmejian challenges were not cited as a problem with most working on the bill. Partisanship, however, was clearly a challenge to passage for AB 2823 as Republicans clearly demonstrated reservations about the bill even in its final version. Appendix C outlines the votes and party affiliation for votes in the Senate and Assembly floors. A majority of Republicans were adamantly opposed to the passage of the bill but did not have a majority in either the Senate or Assembly to prevent passage. Opposition from the bill came from Republicans, the California Grocers Association, the Cosmetic Toiletry and Fragrance Association, and Proctor and Gamble. The opposition raised three primary points of concern.

First, Republicans suggested that a "bounty hunter" provision, defined as an "awesome and powerful incentive to a government bureaucrat," existed because the CDFA would receive direct financial benefit from fees and penalties collected through enforcement activities by CACs (Assembly Republican Committee 2002; Rogers 2002). The rules establishing fees and penalties were law when AB 2832 went up for debate but this still did not deter criticism. As a related and second point of opposition, Republicans also referred to the fees and penalties as "draconian measures" (Rogers 2002). The California Grocers Association was against the proposed fees for handling operations and retailers (Brown 2002a). In general, the new fee structure was widely supported by the organic industry (Strom-Martin 2002a). Eventually, a compromise was made to strike the definition of retailer, which was argued as inconsistent with national law, and to create a $100 flat fee per store for retailers (Green 2002).

A third and significant dispute over AB 2823 was over new regulations regarding cosmetics. In an eleventh hour attempt to derail the bill,[18] Proctor and Gamble along with the Cosmetic Toiletry and Fragrance Association objected to regulations covering organically labeled cosmetics. Cosmetics are not edible items and, therefore, not considered for regulation at that time due to jurisdictional concerns (Lambert 2002), but the NOP did allow for oversight of organically labeled cosmetics requiring a minimum of 70 percent organic content to be labeled as such (Strom-Martin 2002b).[19] It was believed by the opposition that regulatory language covering organically labeled cosmetics was removed during the June 11th amendments because cosmetics were not regulated by the NOP (Livingston 2002a and 2002b). Yet, language regarding organic cosmetics was reinserted with amendments on August 12th and August 20th (Livingston 2002b). Indeed, while certain aspects regulating cosmetics were removed with the June 11th amendments, regulatory language still existed that allowed for oversight of organically labeled cosmetics (AB 2823 Amended 6/11/2002). The August amendments reinserted the minimal standards requiring a minimum of 70 percent organic content in cosmetics before a company could label it as organic.

There was more support voiced for the bill's inclusion of cosmetic regulations. Letters were received from the Environmental Health Network, Nub Circus, Organic Ingredients Inc., Alto Ingredients, Avalon Natural Products, and several individuals in support of maintaining regulations. In the personal letters, individuals expressed gratitude toward Strom-Martin for not succumbing the pressure from Proctor and Gamble and the Cosmetic Toiletries and Fragrance Association (Concerned Citizen 1 2002; Concerned Citizen 2 2002; Concerned Citizen 3 2002; Egide 2002; Peck 2002; Stern 2002; Wilkie 2002).[20] Amendments were not made to accommodate the eleventh hour opposition, and organic cosmetics would be regulated under the California Organic Products Act of 2003.

Amending Adoptions of the 2000s

In the eight years following the passage of the California Organic Products Act of 2003, five legislative bills amended California's organic policy. Amendments ranged from improving enforcement, clarifying expectations of handlers and producers, creating standards for organic fertilizer, and banning the labeling of seafood and aquaculture as organic. Table 4.7 provides a brief summary of each bill.

Most of the modifications were non-controversial. The 2002, 2009, and 2010 bills represent efforts to improve implementation and improve consumer confidence. For example, the 2009 AB 856 creates specific language to identify organic fertilizers and to further specify appropriate practices. The CDFA had been investigating the adulteration of organic fertilizer within the state and found non-organic fertilizer being sold to unknowing organic farmers (Legislative Counsel 2009). Likewise, the 2010 AB 2612 aimed to streamline the registration process, which would improve confidence in the market and law.

SB 730 proposed and adopted in 2005 was the only bill to truly be controversial. The introduction of SB 730 was "in response to concerns that have been raised by several consumers and environmental advocacy groups questioning the truthful and meaningful labeling of organic seafood" (Office of Assemblywoman Jackie Speier 2005).[21] A study conducted by the University of New York at Albany's Institute for Health and the Environment raised concern after results indicated high levels of contaminants including PCBs, dioxins, and banned pesticides in farm-raised fish. At the time, the NOP program did not have established standards for farm-raised or wild-caught aquaculture, fish, or seafood products. In January 2005, the NOP began the

Table 4.7. Bills Amending the California Organic Products Act of 2003

Year	*Bill*	*Summary*
2003	AB 776	The bill clarifies confusion regarding the first-time registration fee of a minimum of $75 for the smallest producers. Scheduled fees are clarified in Section 46013.1 (Legislative Counsel 2003). It also provides more specific language regarding physical description of producer operations (Cal. Food & Ag. Code § 46013.1).
2005	SB 730	The bill prohibits aquaculture products from being labeled, represented or sold as organic until formal standards are developed at the federal level (Cal. Hlth & S. Code § 110827)
2009	AB 856	The bill establishes standards in conjunction with the California Organic Program, California Organic Foods Act of 1990, and California Organic Products Act of 2003 to clearly define organic input material as it relates to organic fertilizer (Legislative Counsel 2009; Cal. Food & Ag. Code § 14528).
2010	AB 2612	The bill clarifies alternatives to the advisory board, adds a definition for an exempt handler, allows for the adoption of an online registration system, and make other technical changes (Legislative Counsel 2010a; Cal. Hlth & S. Code § 110810; Cal. Food & Ag. Code § 46004.1; 46013; 46014.1).
2010	AB 2686	The bill increases the ability for CACs to enforce the law by allowing CACs to go to the District Attorney for criminal prosecution or have County Counsel file for an injunction (Legislative Counsel 2010b; Cal. Food & Ag. Code § 46017).

Source: Data compiled by author.

formulation for two task forces that would develop standards for wild-caught and farm-raised aquatic animals (AMS, n.d.). While standards were being developed, NOP permitted seafood to be labeled with any non-USDA organic label claim (Office of Assemblywoman Jackie Speier 2005). This meant that seafood containing toxic contaminants and prohibited substances could be sold on the U.S. market as organic. Due to federal inaction, the bill was introduced under "the author's hope that SB 730 will set a valuable precedent for the USDA to prohibit the use of the organic label on fish and seafood products until it establishes organic certification standards" (Office of Assemblywoman Jackie Speier 2005).

Opposition to the bill was primarily focused on the introduced version of the bill. As introduced on February 22, 2005, SB 730 prohibited all aquaculture products sold as organic from containing chemicals known to cause cancer or reproductive toxicity. The California Aquaculture Association, among other organizations, claimed the bill "set a precedent that is contrary to existing organic law" (Legislative Counsel 2005). Many Proposition 65 substances occur naturally in the environment and, according to the California Aquaculture Association, unfairly singles out aquaculture products compared to other organic products that may come in contact with naturally occurring carcinogens and toxins.

The Senate passed the final amended version on April 19, 2005. However, there still remained opposition from a number of seafood companies and Republicans, who predominantly voted against the bill on the Senate and Assembly floors.[22] Republicans argued that the private industry and third-party certifiers are the best method for establishing appropriate standards (Assembly Republican Committee 2005). In a letter to Republican governor Arnold Schwarzenegger, Assembly Republican leader Kevin McCarthy (2005) urged the governor to veto the bill on the grounds it "is an overly-broad attempt to allegedly protect consumers from the dangers of farm-raised aquaculture that is labeled 'organic.'" Governor Schwarzenegger did have a veto message included in the CDFA enrolled bill report on August 31, but SB 730 was eventually chaptered into law October 7, 2005.

EXPLAINING ADOPTION

California's organic legislation is an extensive saga for how organic food and agriculture markets came to be regulated by states. Since 1979, a diverse set of factors influenced adoption and reinvention of organic law and policy in the state. Organic farmers primarily drove early legislative efforts via CCOF, who were skeptical of involving government but saw the necessity given changing market conditions. Reinvention in the 1990s and 2000s were also driven by CCOF and organized organic interest groups, but additional factors influenced adoption decisions such as partisan politics, federal adoption, regional pressures, increased salience, and a rapidly growing market. By the turn of the century, the role of CCOF appeared to diminish as a more diverse set of industry interests emerged. Partisan politics, federal implementation, salience, and market factors also continued to influence adoption behavior after the adoption of the California Organic Products Act of 2003. The findings of the case confirm and elaborate upon the statistical results discussed in chapter 3.

Internal Conditions

In the case of California, several internal conditions influenced adoption decisions the most. First and foremost, the role of third-party certification programs and organized organic grassroots groups were highly instrumental in orchestrating the development and passage of policy. CCOF played a significant role in pushing for and writing the 1979, 1982, and 1990 organic bills. Policy was developed in California because organic farmers sought out government intervention to regulate the market. The design was also the result of preferences from organic farmers to maintain the farmer-on-farmer certification (Lipson 2013; Cantisano 2008, 114). As Amigo Bob Cantisano recalls the farmer-on-farmer inspection model was developed and worked well for CCOF and was then used as the model for the Organic Foods Act of 1979 (Cantinsano 2008, 114). The 1979 law was the first organic legislation adoption and established a precedent for a third-party-certification system within the state.

The California Organic Foods Act of 1990 and the California Foods Act of 2003 would maintain and elaborate upon this preference. Under the 1990 bill, certification would not be mandated for all organically labeled products. However, the bill did provide some teeth for enforcement, a desire of CCOF and the organic community. Perhaps an oversight, the organic community left organic certification voluntary without the state actively having to enforce the 1979 law. As Amigo Bob Cantisano reflects, "That was maybe our mistake at that time. We decided in '79 that it would be voluntary to become organic. There was no state mandate. You didn't have to go get registered. You didn't have to go get certified" (Cantisano 2008, 114). The 1990 bill would require registration of producers to identify organic claims. By the time the 2003 bill passed, the role of CCOF appears to have diminished in strength as a larger and more diverse set of players began to enter organic policy debates. However, the third-party certification scheme remained intact in California and was the model used at the federal level.

CCOF's impact on policy development is undeniable. Two key reasons can explain the dominant role of the organization. First, CCOF was not only the only organic certifier in the state of California for the 1970s and much of the 1980s, but also acted as the de facto trade association (Epstein 2013). CCOF worked laboriously in educating and advocating for organic agriculture for much of the 1970s and 1980s. In describing itself today, CCOF identifies itself as a "full-service organic certification agency and trade association passionate about being the leading voice for organic, certifying, educating, advocating, and promoting organic" (CCOF 2013b). The standards and models set forth by the organization have a profound impact on organic agriculture in the United States and abroad.

A second explanation for the dominant role of CCOF in developing standards relates to its capacity as the de facto trade association. In the 1980s, CCOF also served as the de facto enforcer of California's organic law. In oral history interview, Grant Brians stated,

> CCOF really did do due diligence. And every single complaint that went into the CDFA [California Department of Food and Agriculture] during the 1980s,

> after the law was put into place, every single complaint that came from a certification organization or an industry course, came from CCOF, every single one. There were a number of blatant, blatant violators who were caught by people in the industry. They knew that CCOF was willing to follow up on it, even when CDFA wasn't. (Brians 2007, 51)

Brian was suggesting the CCOF ensured that fraudulent organic claims were investigated and enforced. The case of the carrot caper, where Scrowcroft contacted the *San Jose Mercury*, is evidence of CCOF intervening and demonstrating a clear lack of enforcement by state government. Such inaction of California government to prevent fraud would lead to the demands from CCOF for amending legislation in 1990. CCOF could not enforce the law alone.

A second internal characteristic to California that influenced adoption was the presence of an organic market and specialty crop production. While the influence of the organic market could not statistically be examined, qualitative evidence suggests an increasing interest in organic production over time. Scrowcroft and Lipson each indicated a sharp increase of interest in organic farming after the alar scare. Furthermore, the significance of the apple maggot eradication incident in 1988 demonstrates the importance of fruit production and concerns regarding organic farming rights in the state (Scrowcroft 2007, 47; Lipson 2007, 13; Lipson 2013). This increase in interest coincides with the explosive growth in production and sales in the 1990s. Comparatively, the growth of the organic market is similar to the emergence and growth of the specialty crop market starting at the end of the nineteenth century. The emergence and growth of each of these sectors are remarkable and created conditions, specifically a strong organic market anchored strongly in specialty crops, within the state that led to increased pressures to adopt regulation for the market.

Political conditions, a third set of internal characteristics, also played an important role in the adoption of organic legislation in California. The political culture of California is labeled as a moralistic dominant culture with a strong individualistic strain according to Elazar (1994, 284), which is reflective of the 3.55 rating Sharkansky's numeric scale of Elazar's political culture (1969). This type of political culture may have been an undercurrent to the movement in the Central Coast and Central Valley that pushed for organic regulations. It could even be suggested that Santa Cruz, or more specifically the University of California-Santa Cruz, served as the epicenter for the movement. Assemblyman Sam Farr (2007, 18; 2013) even states the organic movement started in Santa Cruz, California, where health and well-being are a significant community concern. He describes Santa Cruz as a very progressive, forward-thinking community particularly after the establishment of the university (Farr 2013). While not representative of all

California communities, the political attitude presented in Santa Cruz and even the Central Coast and Valley would create fertile grounds for a bill regulating the organic market. Specifically, the political culture could accept consumer protection laws (e.g., labeling laws) while also seeking to protect the individualistic interests of organic farmers in the market.

A secondary political influence is the role of partisanship in state government. Democratic legislative members sponsored a majority of bills, but this may have been the result of where their district was located (e.g., in the Central Coast or Central Valley) rather than indication of one political party support organic regulations over another. Only four bills clearly had partisanship play with a clear and evident role in the process or outcome. For each bill, the influence of partisanship didn't appear to be over the issue of organic foods in general but rather focusing on the financial and regulatory burden aspects of the law. Republican Governor Deukmejian created a tremendous amount of pressure for supporters of the California Organic Foods Act of 1990 (Farr 2007, 16; 2013). The 1990 law could not increase the size of government and needed to create its own revenue stream. In part, Republican opposition to the 1990 bill may have been the result of a cavalier attitude toward the potential of the organic market. At the time, there was no indication for how important the bill would be and how quickly the organic market would grow (Farr 2007, 16–17).

As for the other three bills, the issue of financial sustainability of the program was the cause for the 1992 bill veto by Republican governor Pete Wilson and contributed to part of the Republican opposition to the California Organic Products Act of 2003 (Legislative Counsel 1992). Although the organic program had existed in modern form since 1990, Republicans in the 2002 legislative session voiced opposition to the increases in fees and administrative power to collect fees and penalties. The 2005 bill departed from the financial partisan divide and instead was grounded in federalism and the appropriate role for government. Republican opposition clearly objected to the prohibition of organic seafood labeling and presented two different arguments for why the bill should not pass. The first argument was grounded in the idea that private organic certifiers are best equipped to make the judgment of what constitutes organic seafood. The second argument was grounded on the precedents and role of the NOP in developing standards. It could be argued that the role of partisan politics in the case of California is symbolic of party polarization in the United States.

External Conditions

Horizontal and vertical diffusion dimensions also influenced adoption decisions and the design of the bills. Many of the bills had clear evidence of a mixture of both horizontal and vertical dimensions of influence. Some of the influences were based within an elaborate organic network situated in the West and Northeast while other influences originated from state and federal legislation and rule making. Disentangling the precise influence of each factor presents a challenge. As Lipson notes, "[It's] hard to say what the specific influence back and forth might have been" (Lipson 2013).

Horizontal Diffusion Dynamics

Horizontal diffusion occurred both within the region and at a cross-national level. The regional diffusion patterns can be likened more to "cross pollination" rather than clear mimicking in adoption decisions (Lipson 2013). The 1979 bill was based on administrative rules passed in Oregon in 1973. During the adoption process of the California Foods Act of 1990, CCOF members worked in the regional organic organization WACO to standardize practices. The three-year organic transitionary period adopted in 1995 did adhere to the WACO agreement. Members from the Northeast Organic Farming Association chapters also attended WACO meetings to develop a cross-national standardization and normalizing of organic practices.

After 1990, adoption decisions were influenced less by regional activity and more by a national interaction and cross-pollination of concerns and ideas. This national network was prominent in the 1990s but did not become overtly evident as an influence with other amending legislative adoptions in California. The inclusion of the Northeast Organic Farming Association marks how certain states, regions, or hubs of innovation may influence each other. As pointed out by several interview participants, the West, the Northeast, and upper-Midwest regions are really where organic policy is supported and was developed since the 1970s (Lipson 2013; Sustainable Food Advocate 2013). The Midwestern network was not discussed to the depth or extent as the Northeastern network in the case of California, but members from midwestern organizations were a part of the larger national network and discussion.

Vertical Diffusion Dynamics

The effects of vertical diffusion are consistent throughout California's organic legislation history. Approximately one-third of all California adoptions were in response to federal activity in some capacity. The relationship between California adoption and federal activity is likened to a give-and-take scenario. On one hand, California policy makers were respectful of federal supremacy in the policy area. On the other hand, California policy makers lacked faith in the passage of federal law and development of federal standards. Federal adoption did not appear to play as large of an influence when compared to implementation of the NOP or even the Federal Trade Commission's early attempt to develop an organic definition rule.

The influence of federal activity on the state's decisions began in 1979. Policy makers were aware of the Federal Trade Commission developing a definition of organic. Consideration was made as to how California law could potentially conflict with federal law and regulations. Adoptions from 1993 to 1996 included provisions that were in anticipation for when federal standards would be implemented. For example, the 1993 bill was the third amending adoption to have extended the deadline for developing a prohibited substance list in accordance with federal standards. After waiting two years for federal implementation, the 1993 bill extended the deadline to 1996. In turn, continual federal delays led to California adopting legislation to extend the prohibited substance list deadline to when federal rules are adopted. In 2002, California was finally able to adjust its law to comply with the NOP. Statutory language changes within the California Food and Agriculture and Health and Safety codes reflect federal supremacy in developing standards.

Somewhat surprisingly, California was not directly influenced by the passage of the national Organic Foods Production Act in 1990. The development and passage of California and federal law can best be described as operating on parallel trucks (Lipson 2007, 19; 2013). In other words, the development of California's law was not influenced by federal legislation despite both efforts being in response to the same set of conditions. A probable cause for this state-federal relationship hinged on the probability of passage. As Lipson describes, "They both were moving towards completion. There was no guarantee that either one of them would actually, ultimately come to be enacted. So both had to proceed as if the other wasn't even happening. Or at least that was the case in California" (Lipson 2013). If anything, it was California that bared more influence on the federal movement for organic food and agriculture regulation. Lipson, "midwife for the California effort," contacted the U.S. Senate Committee on Agriculture, Nutrition, and Forestry in 1989 and spoke to Kathleen Merrigan, a legislative staff member for Senator Patrick Leahy (Merrigan 2012; Lipson 2013).[23] Merrigan, who would later work to develop NOP standards and service the USDA Deputy Secretary of Agriculture, then worked to push the issue onto the agenda by working with organic farmers, industry leaders, and environmental activists (Merrigan 2012). Merrigan would become midwife of the national organic standards and was an instrumental figure in the creation and development of federal law.

In 2005 and 2009, California policy makers would take the lead on organic regulations. Delayed federal response to new issues emerging from an expanding and growing market led to California responding with its own policy.

The 2009 bill came after a couple of years of investigating organic fertilizers and acceptable organic fertilizer practices. Two major cases of organic fertilizer fraud in California raised the alarm (AMS 2013). The NOP did issue a final rule on organic fertilizers in 2009, but the rule would not go into effect until 2011 (NOP 2009). This is similar to the regulation of organically labeled aquaculture in 2005. The NOP's decision to permit aquaculture to be labeled as organic until rules were developed was questionable to many. California reacted sooner to address the problem of organically labeled aquaculture to protect consumer confidence. Democratic lawmakers saw federal law as unacceptable and moved to prohibit the labeling of aquaculture in the state until the NOP developed standards. In sum, California influenced and moved forward federal organic policy as much as federal supremacy influenced California's organic food and agriculture legislation development.

Salience

The role of salience in the case of California is undeniable. A combination of focusing events and increasing coverage of organic food and agriculture fermented conditions ripe for the proposal and push for legislation. Attention to the watering-down of the term natural within the organic community compelled action to develop rules to protect organic labeling from a similar fate in the late 1970s. The carrot caper incident and alar scare in the late 1980s were two significant focusing events that were part of a larger rising tide of media attention toward organic fraud and alarm of conventional agricultural practices. Each event stimulated interest in promoting organic agriculture, preventing fraud in the market, and maintaining consumer confidence in labeling claims. In a Legislative Counsel Digest for 1990 AB 2012, several comments made about the public and media attention focused on organic foods, including the Alar scare, growth in organic production, and consumer preference for organic products. Similar references were made in 2005 and 2009 with organic seafood and organic fertilizers. The 2005 SB 730 digest referred to NYU-Albany Institute for Health and the Environment study on contaminants in wild and farm-raised salmon (Office of Assemblywoman Jackie Speier 2005). Likewise, 2009 AB 856 was in response to high-profile cases of organic fertilizer fraud and a CDFA investigation into the issue (NOP 2011). What can be inferred from each example is that salience matters particularly in conjunction with specific focusing events especially when accounting for a lagged response in earlier adoption cases. While specific focusing events were not accounted for in the statistical model, it is clear that increasing attention to the issue and problems in the organic market clearly led to legislative adoption in at least four of the eighteen legislative adoption decisions.

CONCLUDING CALIFORNIA

The development of organic policy in California's state legislature is remarkable. The complexity of how variables influenced adoption decisions leads to an intricate story and explanation for how organic policy developed in the state. The influence of CCOF, a third-party certifier and advocacy group, is one predominant factor that explains adoption behavior. Issue salience, market growth and expansion, regional factors, federal activity or inactivity, and political considerations, such as an underlying political culture perspective, were also influential. The factor of time determines how each variable uniquely factors into adoption. The initial adoption, which made California an early-adopter, was spurred by CCOF and the political culture of the Central Coast and Central Valley. The 1990 reform adoptions were also influenced by CCOF, but the role of salience, regional influence, market growth, and partisan politics also influenced the development of the policy to where it was acceptable for passage. Federal activity, salience, partisan politics, and an expanding and growing market influenced the 2002 reform adoption and subsequent amendments. The difference between initial and later amending adoptions is evident in how the issue of organic food and agriculture changed over time and the number of actors in the policy arena expanded. The story of organic regulatory legislation in California points to a long history of how nongovernmental actors influenced the emergence of organic regulatory standards in the United States.

NOTES

1. The percentage of California's total GDP resulting from agricultural industry from 1976 to 2010 ranges from a high of 3.1 percent in 1979 to a low of 1.3 percent in 2001 and 2006. In 2010, the percentage of California's GDP resulting from agriculture was 1.6 percent. These percentages were calculated using USDA's Ag Census Data.

2. The Small Farm Program was created in 1976 and was the predecessor of the Small Farm Center (Christensen et al. 1990). In 2010, the Small Farm Center closed and became the Small Farm Program (UCSFP 2014).

3. According to Martin and Olmstead (1985), "Mechanization research is construed to include the development of machinery, crop varieties, chemical herbicides, growth regulators, and laborsaving methods of handling, transporting, and processing crops."

4. The Farm is a separate operating entity from the Garden. The Farm Center was built under the premise of a class and founded in 1972 (Martin 2008, 17; CASFS 2010b).

5. In the summer of 1985, two thousand people were sickened after eating California-grown watermelons illegally contaminated with Aldicarb, a pesticide used to kill pests on cotton, citrus, and potatoes among other food crops (Cone 2010).

6. On June 6, 1978, California voters passed Proposition 13. As part of the larger anti-tax movement in the country, Proposition 13 decreased property taxes to 1 percent and limited the growth of future assessments. It was believed, by some, that the passage of Proposition 13 would lead to cuts in government spending and increase unemployment rates (Glyn and Drenkard 2013).

7. Sy Weisman and Stuart's Fishman's North Coast chapter was the Farmers Organic Group of Sonoma County, a marketing co-op (CCOF 1988). Fishman was a produce buyer for Rainbow Groceries, a San Francisco co-op grocery store, and was reputable for scrutinizing organic operations. The term "Stu Fished" was a phrase coined to describe thorough scrutiny of an organic operation (Burroughs 2004).

8. Assemblyman Gage's office received opposition letters from others. One example is William Jarvis (1979), representing the National Council Against Health Fraud, argued that the bill falls short for consumers and should not legitimize terms that would be misleading in the long run. Another example is the California Association of Tobacco and Candy Distributors raising concern about having to adjust marketing of wild cherry cough drops to comply with regulatory language regarding the term wild (Fishman and Fishman 1979).

9. The medfly, or Mediterranean fruit fly, is an invasive species to North America that can cause extensive damage to a number of fruit crops. Considered a pest, the apple maggot fly primarily infests apple crops but can also infests other fruit crops including cherries, plums, prunes, apricots, nectarines, peaches, blueberries, and pears. The exception was made due to the medfly infestation occurring in San Mateo, Alameda, and Santa Clara counties among other areas (Hagen, Allen, and Tassan 1981).

10. Lipson (2007, 20) also noted the Chilean grape scare, where imported grapes were tainted with cyanide, in 1989, as a significant event.

11. Effort was made to create a conducive and welcoming environment for a wide variety of participants to the extent of Epstein requesting decorum at meetings (Epstein 1989a). Meeting participants include but are not limited to CCOF, CDFA, CDHS, California Department of Finance, California Cattlemen's Association, California Grocers Association, Butte County Farm Bureau, Santa Cruz, County Farm Bureau, California Grain and Feed Association, California League of Food Processors, Western Growers Association, and California Farm Bureau.

12. As Epstein (2013) indicated as an appropriate footnote, California's legislative session acts as a two-year legislative session. If bills are introduced in the first year of the two-year session window, such as the 1989–1990 session, the bill will continue on the books during the second year session.

13. Another elected official that represented a challenge was Assemblyman Norman Waters, a Democrat and chair of the Assembly Agriculture Committee. Waters was identified as, "an established, typical state politician, big friends with big ag," and likely to create problems with reform, particularly after organic growers expressed disgust with the 1988 mandatory spraying for medflies (Lipson 2007, 16).

14. Associated with the revenue concern is concern about where collected fees would be deposited. Earlier versions of the bill directed collected fees to be deposited in the Food and Agriculture Fund and the Health Safety Fund, which did not exist.

An adjustment was made for fees to be deposited into the California General Fund (California Department of Finance [CDF] 1990).

15. Collection of penalties and administrative charges were also additional sources of funding that would fund, but not significantly, the program. Registration fees were expected to garner around $400,000 initially. Estimated cost for CDFA and CACs were estimated at $425,000 and $1.2 million for CDHS (Lipson 1990).

16. Interested and supportive parties throughout the process included CCOF, the California Farm Bureau, the Environmental Health Network, the California Council for International Trade, Bayliss Ranch, Lundberg Farms, California Natural Products, Avalon Natural Products, Organic Ingredients Inc., Nub Circus, Alto Ingredients, Tikvah Company, and twenty-six individuals including consumers. California Grocers Association, Proctor and Gamble, and the Cosmetic Toiletries and Fragrance Association voiced opposition.

17. Eventually a COPA task force, led by Gay Timmons, developed in 2000 and comprised of approximately sixty-five companies and individuals to develop the proposed amendments (Timmons 2002a and 2002b).

18. Opposition to the bill's regulation of cosmetics was not raised until August 2002, approximately two weeks before passage on the Senate and Assembly floors. A formal letter of opposition was not received until August 26, 2002 (Strom-Martin 2002b).

19. For information about federal oversight of organic cosmetics see http://www.fda.gov/Cosmetics/Labeling/Claims/ucm203078.htm#Does_FDA.

20. One very personal letter indicated disgust with Clairol Herbal Essence Shampoo, which was labeled as "totally organic." The product was reported to cause several skin and breathing conditions for that individual (Concerned Citizen 1 2002).

21. The American Federation of State, County, and Municipal Employee California Public Interest Research Group, the Center for Food Safety, the Coastal Alliance for Aquaculture Reform, Consumers Union, the Organic Consumers Association, and other public citizens expressed support for the bill (Assembly Republican Committee 2005).

22. Opposing seafood companies include 8th Sea Organic Seafood Company, Emerald Organics LLC, and Horizon Organic Seafood Company Inc. (Assembly Republican Committee 2005).

23. Kathleen Merrigan references initial contact with Lipson in 1988 (Merrigan 2012).

Chapter Five

Organic in Vermont

A Critical Mass Adopter

Organic regulatory policy first appeared in Vermont in 1990. Like California, the state had a long history of organic agricultural production, but government involvement was delayed. In fact, the state was a regional laggard behind New York, Connecticut, Maine, New Hampshire, and Massachusetts in the adoption of organic regulatory policy. Nationally, Vermont's state legislature adopted at the end of a swelling critical mass, adopting in the same year as Congress passed OFPA. The delay of action can be traced back to grassroots organizations representing organic farmers in the state who were skeptical of the government developing organic standards. Vermont experienced the same back-to-land and counterculture movements in the 1960s and 1970s as California; yet, the key actors operating organic organizations in the state did not pursue state policies to protect their livelihood until a critical mass of interest emerged.

The case of Vermont is one that illuminates how the groundswell of attention to organic agriculture led to action in states from 1985–1990. The increasing interaction among organic farmers and certifiers nationally and regionally contributed to elevated attention levels to the problem of regulating the organic market. The policy environment of Vermont permits an evaluation of how dissenting yet supportive organic perspectives led to a unique approach to regulating the market. The producer-oriented approach competes with the more stringent consumer-approach developed in California. In addition, the story of Vermont's policy also demonstrates the significance of federal supremacy in changing the trajectory of state policy objectives.

In what follows, an overview of the socio-economic and political conditions of Vermont is reviewed, followed by a historical discussion of Vermont's agricultural industry. Next, an overview of the organic agricultural market is provided and the role of the Northeast Organic Farming Association is discussed. Subsequently, the chapter then turns to discussing the adoption of the 1990 Organic Farm Products Act, a 1998 joint resolution adopted in light of the 1997 NOP proposed rules, and the termination of statutory language in 2003. Finally, the chapter concludes with an evaluation of significant determinants.

VERMONT POLITICAL AND SOCIO-ECONOMIC CONDITIONS

The Green Mountain State is a culturally distinctive case. Vermont is a largely rural state with a small economy and a political environment open to politically progressive ideas (see appendix D for population density and income). The state economy is relatively diverse for its size with agriculture being a proud industry within the state. Vermont possesses a strong agritourism and alternative agricultural market compared to other states. Politically, the culture and environment within the state may support interest in sustainable and alternative agriculture markets. State residents are more liberal but see a particular role for government in society.

Politics in Vermont

Vermont has been described as one of the more independent and liberally progressive states in the country. Perhaps marching to the beat of its own drummer, Vermont has been recently ranked as the fourth most liberal state, behind California, New York, and New Jersey, but has a mixture of Republicans, Democrats, Independents, and Progressive party members that serve at the state and federal level (Gray 2012).[1] The once Republican stronghold became more liberal in the 1960s as "an influx of urban refugees and hippie escapists," predominantly from Boston and New York City, fled to greener pastures. According to Johnson and Nelson (as referenced by Cohen 2012), Vermont attracted more liberal, environmentally concerned professionals while more conservative city-migrants moved to the neighboring state of New Hampshire. The mix of new immigrants with native-Vermonters has led to diverse party representation in a state where individuals are more likely to describe ideological affiliation rather than party affiliation (Cohen 2012). However, Vermont voters do not favor Republicans in national elections. For example, in the last four presidential elections, voters from Vermont have predominantly favored Democrats. In 2008 and 2010, Obama won in all fourteen counties (see appendix D).

Vermont is politically similar to other states in the Northeast region (see appendix D). New Hampshire is the only other state to have a historical track record of partisan control by both Democrats and Republicans. All other regional states vary between Democrat and split control. Despite slight partisan control differences, the political culture is near average for the region. Northeastern states fall between moralistic and individualistic cultures with an average rating of 2.89 on Sharkansky's scale of Elazar's politics culture. Vermont's political culture is ranked as 2.33, the same as Maine and New Hampshire, and is identified as a moralistic culture with an individualistic strain. Similar to California, Vermont's culture is one that possesses a healthy dose of skepticism of government and emphasizes individual rights, but the state has a less professionalized state legislature. The level of professionalism is steady over time with peaks in 1986 (rating of .145) and 2003 (rating of .144). After 2004, the level of professionalism declined to .104, the lowest since 1976. For the region, this level of professionalism is slightly lower than average. New York's legislature is the most professionalized (rating of .627 in 2010) whereas New Hampshire's legislature is the least (rating of .032 in 2010).

The Agriculture Industry in Vermont

Vermont is consistently ranked as having the smallest state economy in the country (BEA 2015a). The average GDP from 2009–2014 is 27.6 billion (BEA 2015b). The top-five private industries in the state include retail, finance, professional services, education services, and manufacturing (BEA 2015a and 2015b). Agriculture is not a top industry with the average agricultural industry GDP of $371 million from 2008–2014. Yet despite the industry's overall contribution to the state economic activities, agriculture has a long and proud history in the state of Vermont.

The current state agricultural market in Vermont boasts more farmers markets, farm stands, and community-supported agricultural ventures per capita than any other state in the nation (USDA Ag Census 2012). Agriculture is a significant industry within the state as it serves as a hub for dairy, maple, and specialty crop production such as apples. Dairy, including milk and other products like cheese, is by far the highest market value commodity with approximately three-fourths of total Vermont farm product sales resulting from dairy (USDA Ag Census 2012). Although a smaller percentage of total market value sales, Vermont is the leading producer of maple syrup in the country supplying 1.39 million gallons or 41 percent of U.S. syrup in 2014 (NASS 2015a). Fruits and vegetables sales generate as much revenue as maple syrup with 2012 figures for fruit and vegetable sales reaching $28,009,000 compared to $26,625,000 from maple syrup (NASS 2013). In addition to commodity sales, agritourism is also a significant source of state revenue as visitors may seek to visit a number of iconic Green Mountain farms and factories of such brands as Cabot Creamery, Ben and Jerry's Ice Cream, and Maple Grove Farms.

The relationship between agriculture and the state of Vermont is deep. For some, the development of the state as a whole is inextricably linked to historical agricultural production trends. Geographically, farmers in the state of Vermont face a number of challenges. The Northeastern region has been classified as an area difficult to farm. It is described as hilly, stony, and having thin-soil (Haystead and Fite 1955). Despite poor growing conditions, early Vermont settlers, predominately English Europeans, managed small-scale subsistence farms. Eventually, major exports from the pioneer region included potatoes, grains, cattle, maple products, potash, and pearl ash (Bidwell and Falconer 1941; NPS, n.d.; PLACE, n.d.). Maple sugar production was particularly significant for Vermont; descriptive accounts of pioneers identify approximately two-thirds of all families in the state participating in the industry (Bidwell and Falconer 1941, 80). Most of the maple sugar would later be exported to nearby states including New York and Massachusetts.

By the nineteenth century, the Vermont landscape began to change. Sheep farming was particularly advantageous for farmers due to the rocky terrain. The opening of the Champlain and Erie Canals in the 1820s strengthened the state's merino wool industry only to later decline after the elimination of a protective tariff in 1846 and the expansion of the railroad system in the Reconstruction Era (Bezio 2009; NPS, n.d.). Farmers would seek to diversify their operations in addition to moving into dairy production. Butter, cheese, and milk products began export into neighboring states in the mid-nineteenth century. By the turn of the century, dairy production and exports increased as market and transportation conditions, such as refrigeration, improved. The dairy industry has remained a dominant agricultural sector even accounting for changes during the latter half of the twentieth century.

The 1980s marked a notable shift in state agriculture. The number of family farms declined in the state, a trend that was both long-term and partially in response to the farming crises of the mid-1980s. In addition, the land was beginning to shift back into forests or better-developed communities (Rural Vermont 2015).[2] Farms in more mountainous or hilly areas tended to disappear as opposed to those located in valleys or low-lying areas.[3] After surviving the crises of the 1980s, dairy production remained steady as better agricultural practices were introduced and diversification of agricultural production continued (Office of Governor Peter Shumlin 2011; Parsons 2010).[4] Farms in the state still remain small (i.e., under one hundred acres) with a recent increase in the total number of farms (USDA Ag Census 2012). Farms averaging less than fifty acres are where most growth has occurred. Production focus still remains diversified with the state still specializing in dairy, maple sugar, and specialty crop production.

Vermont Organic Agriculture

The emergence of organic agriculture in Vermont is a comparable story to California. The same spirit that inspired organic farmers in northern California appeared in Vermont. The back-to-the-land movement and heighten environmental awareness drove an interest among farmers to develop a deeper and more meaningful connection with the land. Inspired by such philosophical thinkers as Al Krebs, Ralph Borsodi, and Wendall Berry, early organic famers in Vermont sought an emphasis on locally grown foods that was provided in harmony with what nature offered. Unlike California, however, Vermont did not have a university presence in cultivating interest and support for organic techniques until the 1990s (Former NOFA-VT Staff Member 2015; NOFA-VT Staff Member 2015). Instead, organic farmers became enlightened through a network in the Northeast and developed an association spanning Vermont and New Hampshire.

In the early to mid-twentieth century, a number of farmers in the Northeast were enlightened by a number of foundational European organic innovators. The most foundational and well-known efforts were located in Pennsylvania, located just outside the Northeast region in the Mid-Atlantic, including Kimberton Hills Farm, Walnut Acres Farm, and the Rodale Institute. The Rodale Institute is perhaps the most well known outside of the organic community. J. I. Rodale is credited with popularizing organic production techniques through the publication *Organic Farming and Gardening*. Magazine publications began in 1942 and J. I. Rodale would later establish an experimental farm, which inspired a number of articles about organic practices. Issues of *Organic Farming and Gardening* would eventually land in the hands of organic farming pioneers in Vermont.

In 1971, as the story is told, the Northeast Organic Farming Association (NOFA) was founded in Putney, Vermont, by Samuel Kaymen, better known as the man behind Stonyfield Farm, and farmers from both Vermont and New Hampshire (Gershuny 1986).[5] The founders were concerned about an eventual collapse of "the system" and sought to distribute locally grown organic produce in the region through co-ops (Gershuny 1986 and 2016). A number of other organizations in Vermont and the region would also emerge around the same time including the founding of Maine Organic Farmers and Gardeners Association in 1971. The primary emphasis of NOFA in the early years focused on distribution of produce and creating access to farming materials that were more difficult to obtain. Grants helped to sustain the organization's activities, and there were frequent meetings, conferences, and gatherings to move the organization forward. There was a great deal of suspicion from established grocers about the organizers distribution tactic of farmers' markets and co-cops, but NOFA persevered in its efforts to grow local organic foods that could be offered to consumers.

Eventually, NOFA professionalized, branched into different state chapters and the Vermont chapter pursued a certification function under an association called the Vermont Organic Farmers (VOF). Table 5.1 outlines key milestones in the history of NOFA.[6] Interest in certification was minimal in the early years. As former NOFA certification coordinator Grace Gershuny recalls, there was only one interested customer in 1977. Interested parties were somewhat split in seeing the benefit to certification. Fraudulent claims of organic were present, but some farmers saw little value-added in obtaining certification. Instead, developing trusted relationships was a preferred alternative until 1979. When NOFA split into two chapters representing Vermont and New Hampshire, there were several interested parties seeking certification.

Interest in organic farming in the 1970s was marginal and perceived as a fringe practice. Interest in organics would explode in the 1980s from both the supply and demand side of the market. The decade is described as having a heightened sense of awareness marked by an increasingly connected organic communication network regionally and nationally. From 1982 to

Table 5.1. NOFA Chapter Timeline

Year	*Event*
1971	Northeast Organic Farmers Association established with joint effort of membership base in Vermont and New Hampshire
	Maine Organic Farmers and Gardeners Association established
1979	NOFA splits into NOFA-VT and NOFA-NH
1982	NOFA chapters established in Massachusetts, Connecticut and New York*
1985	NOFA chapter established in New Jersey
1989	NOFA changed association name from "Farmers" to "Farming"
1990	NOFA chapter established in Rhode Island

**Some reports suggest that New York may have been established in 1983 instead of 1982.*

Source: Data compiled from Gershuny 1986 and 2016 ; Rawson et al. 2011.

1990, five additional chapters of NOFA would be established and leaders of NOFA and VOF recall national interactions with CCOF, Oregon Tilth, and Washinton Tilth to standardize organic certification practices. In addition, certification of organic operations continued with the Vermont department of agriculture jointly aiding NOFA-VT in certifying operations from 1985–1990 primarily through grant support (Wonnacott 2015). The partnership developed after a major distributor from the Mid-Atlantic came to Vermont seeking certified organic produce to sell in 1984. NOFA-VT strengthened its standards to be similar to the Organic Crop Improvement Association (OCIA) certification rules and then sought help for certifying operations (DiMatteo and Gershuny 2007).[7]

The added value for "certified" organic increased through the remainder of the 1980s and would double in the 1990s (see figures 5.1 and 5.2). In 1987, only seventeen farmers were certified by VOF and by 1993 the total number of producers certified was seventy-eight (NOFA-VT 2015; Wonnacott, n.d.). Today, the organic sector has continued to expand with 513 certified farms operating in the state and 2014 sales reaching $92.1 million (NASS 2015b; USDA 2012). Milk represents the largest portion of sales with $46.43 million. Certified dairy producers drastically increased in 2007 with the sector representing the fastest growing segment of Vermont agriculture since 2000 (Parsons 2015; NOFA-VT 2014).

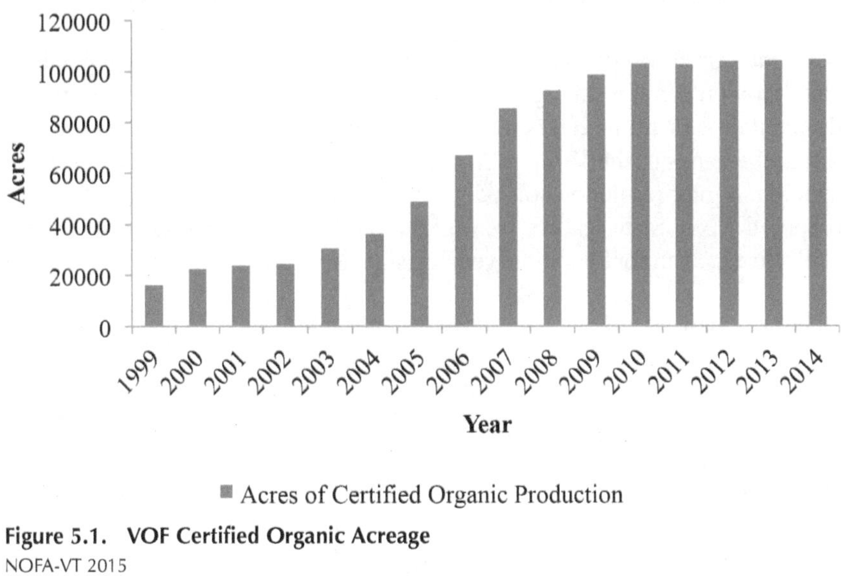

Figure 5.1. VOF Certified Organic Acreage
NOFA-VT 2015

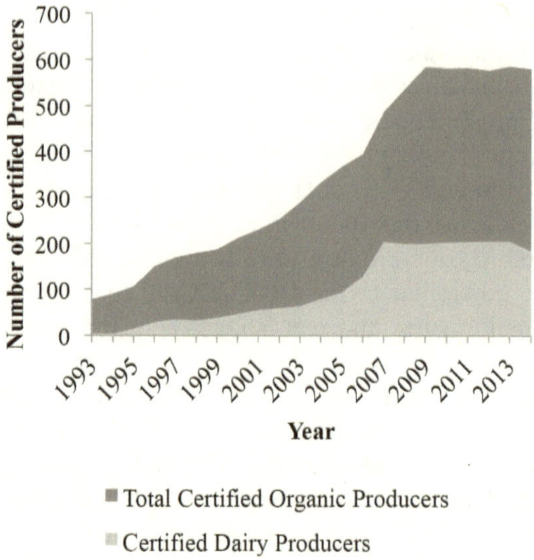

Figure 5.2. VOF Certified Organic Producers
NOFA-VT 2015

A HISTORY OF ORGANIC FOOD AND AGRICULTURE LEGISLATION IN VERMONT

Vermont passed two bills pertaining to organic food and agriculture regulation. One bill would create an organic advisory board in 1990 and the other would dismantle the board in 2003 (see table 5.2). Implementation of the law would rest in the hands of the Vermont Agency of Agriculture, Food, and Markets. In addition, a joint resolution was adopted in 1998 in response to the 1997 NOP proposed rules. State leaders were highly critical of the proposed measures. Collectively, Vermont's short-lived legislative efforts represent a nationally

Table 5.2. Vermont Organic Food and Agriculture Legislation

Year	*Bill*	*Sponsored/Introduced**
1990	S.277 *An Act Relating to Organic Farm Products*	Elizabeth Ready [D]
1998	JRH 167	David Zuckerman [P], Dean Corren [P], Jennifer Nelson [D], and Andrew Snyder [D]
2003	S.11	Sara Branon Kittell [D]

**Political Party – [D] = Democrat; [P] = Progressive; [R] = Republican*

Source: Data compiled by author.

dissenting perspective of government involvement in organic standards and critically aware understanding of other state's implementation efforts.

Vermont's story is explored in three stages. First, the events surrounding the passage of S.277 in 1990 are discussed. Next, the factors motivating the adoption of JRH 167 in 1998 are detailed. JRH 167 was not included as part of the statistical analysis, but the resolution highlights some of the underlying dynamics surrounding a Vermont perspective on organic farming regulations. Finally, a brief synopsis covers the dismantling and termination of the Vermont Organic Advisory Board in 2003.

1990 Organic Farm Products Act

On January 3, 1990, S.277 was read for the first time in the Vermont Senate. Proponents of the bill spent the previous year discussing and negotiating the contents of the Act relating to organic farm products. The bill possessed both regulatory and distributive educational components (see table 5.3). Section 1 of S.277 created an organic advisory board under the agriculture title as 6 V.S.A. § 181. The agriculture commissioner, who was also responsible for the appointment of the other six members, led the seven-member board. The appointed six members, with three of them being organic farmers, were to advise the commissioner in the development of rules for the labeling and certification of organic foods. Complementing the regulatory component was the creation of a sustainable agriculture research and education program codified as 6 V.S.A. § 2321a. The bill was packaged not only to regulate the market but also to provide a platform for educating students and providing support for sustainable agricultural production.

Table 5.3. Summary of S.277/ 6 V.S.A. § 181 and § 2321a

Aspect	*Summary*
Regulatory	Establishes a 7-member Organic Advisory Board to promulgate rules for organic labeling and certification. Led by the Commissioner with six additional seats being appointed. Three seats are reserved for organic farmers. Three seats are designated as "other" (6 V.S.A. § 181).
Educational	Establishes the sustainable agriculture research and education program to promote research and educational opportunities to "enhance the economic, environmental and social sustainability of Vermont agriculture." The commissioner would develop an agreement with the University of Vermont, Vermont Technical College, and an approved member from a sustainable agriculture organization to implement the program (6 V.S.A. § 2321a).

Source: Data compiled by author from state bill and statute information.

The primary sponsor of the bill was Elizabeth Ready. Ready is a Democrat from Addison County that served from 1989–2000. During her tenure, she served eight years on the Senate Agriculture Committee, six years on the Senate Appropriations Committee, six years on the joint-chamber Fiscal Committee, and twelve years on the Natural Resources Committee with four years service as chairwoman. Ready's sponsorship of the bill came after being approached by members of the VOF and NOFA-VT, some of which were her constituents (Former VOF Staff Member 2015). Ready's personal ideology, which is described as being a left-leaning Democrat, and representation of the organic farming community, made her an ideal ally. She began working with VOF and NOFA-VT on crafting a bill shortly after entering office. VOF and NOFA-VT had the support of the Vermont's agriculture department.

In part, the bill was a "truth in advertising" effort to protect consumers (Christensen 1990). However, the bill was also seeking to establish the importance of organic farmers and producers in determining organic standards. Four interrelated factors influenced organic farmers in seeking government-backed regulations. First, the tide of interest in organic foods in the late 1980s impacted the organic farming community in Vermont as much as it did in California. Meryl Streep's appearance on *60 Minutes* in February 1989 was a national focusing event that could not be escaped. (See chapter 4 for more information about Streep's *60 Minutes* appearance.) Rising consumer interest in organic and non-conventional products rapidly elevated in part because of focus on the dangers of conventional farming. Another contributing factor to increased attention was fraud in the marketplace emerged both locally and nationally.

As a second factor, organic advocates in Vermont were keenly aware of multiple cases of fraud in the marketplace including the California carrot caper scandal (for more information see chapter 6). Locally, NOFA-VT and VOF staff members were cautious in dealing with fraud. One particular case surrounded farmer George Crane. Considered the Vermont equivalent to the California carrot caper scandal, Crane was a "borderline grower" stretching the authenticity of his organic labeling claims for several years (Gershuny 2016; Former NOFA-VT Staff Member 2015). The first questionable claim was in 1978 after Gershuny investigated Crane's organic oats crop for an interested buyer. Crane's operations would later come crumbling down after he was caught importing conventionally grown carrots and repackaging them as organic. Crane's fraud was not widely known and was largely kept quiet by organic certifiers and organizations in the state.

A third motivating factor centered on the actions of other states both nationally and regionally. In Senate and House testimony, bill advocates outlined a number of challenges seen in the rising tide of state regulatory activity. Drafters of S.277 were aware of at least twenty-five other states that "either have organic labeling laws on the books or are in the process of drafting them." Vermont advocates were seeking to avoid many of the implementation challenges seen in other states (Gershuny 1990). Texas, Colorado, and Washington were cited as having state-operated certification programs, which Vermont advocates saw as displacing the professional knowledge of organic growers (Gershuny 1990). Additional complaints were lodged against California's and Massachusetts's implementation designs (Stevens 1990). California was noted as having regulations that worked "contrary to the certification process." Massachusetts's administrative rules were developed by the health department, creating strange bedfellows to the dismay of producers. Vermont activists saw other designs, which can be classified as consumer-oriented, a direct threat to the nature of organic farming.

Vermont advocates sought to avoid similar implementation scenarios by designing S.277 to essentially certify the certifiers (Former VOF Staff Member 2015). NOFA-VT and VOF had established an excellent and mutually respectful working relationship with the Vermont Agency of Agriculture, Food, and Markets in the mid-1980s. The administrative support, financial and otherwise, for certification efforts in the state created "the best working relationship between a state agriculture department and NOFA certification committee" (Stevens 1990).[8] The partnership helped cement the importance of producers in setting standards and prevented a co-optation of rule development by larger agricultural interests or less knowledgeable administrators and consumers.

A fourth and final factor converging on Vermont in 1990 were the inevitable actions at the federal level. Shortly after Meryl Streep appeared on *60 Minutes* to talk about Alar, Senator Leahy's (VT-D) congressional staff member, Kathleen Merrigan, visited Montpelier to talk with Leahy's constituents about what would be happening at the federal level to regulate the organic market (Former VOF Staff Member 2015). Leahy would sponsor the Organic Foods Production Act of 1990. However, the collective Vermont response to federal efforts was negative and represented a dissenting opinion nationally. Growers and advocacy groups felt the federal effort would overlook some of the "finer points of organic production" by creating a consumer-focused law rather than a producer-focused law (Former VOF Staff Member 2015). In essence, the federal law would create a regulatory structure that removed the power to define organic practices out of the hands of producers, a more knowledgeable body on appropriate standards, and to government bureaucrats, who did not possess the same sophisticated knowledge-level as growers. The idea of a consumer-oriented approach could permit practices that were at odds with the philosophical nature of organic farming. In essence, the focus of consumer laws is product-driven and focuses on the shelf-label meaning (see Clark 2015). Vermont's law would be process-based and permit for variations in farming practices to accommodate farm type needs.

S.277 was a response to federal efforts (Former VOF Staff Member 2015; NOFA-VT Staff Member 2015). Those at NOFA-VT and VOF felt they had been able to manage reciprocity issues that arose from interstate trade and adequately managed cases of fraud. Vermont organic producers believed the best market management and oversight would come from growers and producers, who are the experts. Correcting the patchwork of state regulations could occur as certifiers worked together to standardize appropriate practices. Unlike the federal law, Vermont's approach would create a floor standard, or a set of minimum requirements for being organic, but it would not create a ceiling and would allow for varying degrees to be more organic. The creation of a board with clear representation of farmers kept the regulatory efforts as a primarily production-focused law.

Amendments and Compromise

S.277 faced very limited opposition. Growers were united behind the bill (Vermont Senate Agriculture Committee 1990). Ideologically, some conservatives were concerned if the bill was necessary, since an individual might not be able to tell the difference between organic and conventionally grown goods. However, this complaint was not a significant deterrent to passage. Instead, funding of implementation efforts, perhaps a non-partisan concern, and interstate trade challenges were behind three significant changes to the bill before passage.

The first modification of S.277 added language to sub-clause (a) of 6 V.S.A. § 181. Nine words were added to the end of the clause that read as follows: "or legal labeling of the sending state or country." The language was added, "to protect Vermont consumers from fraudulent claims on food imported into the state" (Christensen 1990). Vermont's agriculture department noted a concern of imported maple syrup labeled as organic. As a key product for the state, the department preferred the empowerment to investigate the claims of organic and prevent mislabeling the market.

The second modification was the addition of the sustainable agriculture research and education program, which constituted section 2. Proponents included the educational program component to aid market expansion. Specifically, government-backed certification standards would ensure quality and truth in labeling. In turn, the market for organic foods, both within Vermont and in other states, could be expanded to meet consumer demand if supply was available. The education and extension program could help to foster supply to meet to increasing demand (Christensen 1990).

The third and most significant modification to S.277 centered on the creation of the Organic Advisory Board. The bill initially created a state organizational center that would focus on sustainable agricultural production. Except when the bill became under consideration in the Vermont House of Representatives, two concerns were raised about the creation of the sustainable agriculture center (Wonnacott 2015). First, there were underlying concerns that the center would not be significantly different than the USDA Low-Input Sustainable Agriculture (LISA) program and would require funding at some point in the future. Congress appropriated funds in 1988 for the implementation of LISA, which provided support for scientists, farmers, and universities in the area of sustainable agriculture. The northeast was a regional focus of the program with students and faculty from the University of Vermont benefitting from the program. Vermont's center did not require funding for the first year of implementation but a concern was raised about the potential need for funding in the future. As a second and related concern, the center was perceived as "another level of bureaucracy" that could be avoided (Wonnacott 2015). Instead of creating new center modifications to existing programs within the agriculture department it could integrate sustainable agriculture objectives. This would help to avoid a program similar to LISA.

Termination of language creating a center was necessary to garner support in the House of Representatives but proponents still feared the bill would fail. The Vermont Organic Advisory Board was proposed and added to the final bill to further aid in passage and maintain near universal support. The board would represent the interests of the agriculture department and organic producers and distributors to protect consumers from marketplace fraud. With the final amendments, the bill would pass the House and would be signed by the governor on June 4, 1990.

JRH 167—An Action in an Era of Uncertainty

The Organic Advisory Board met a few times after enactment of S.277. Discussions were centered on regulatory concerns such as land-conversion wait periods, certification processes, and general administrative activities. However, board activity began to diminish once it became clear the federal law would eventually go into effect (Former VOF Staff Member 2015; Wonnacott 2015). The state did not reinvent or amend the 1990s policy to accommodate for federal actions. Nevertheless lawmakers still closely watched the activity of the National Organic Standards Board. After the 1997 NOP proposed rules were published, lawmakers were swift to respond to the inadequacy of the federal design through a joint resolution.

Led by Progressive senator David Zuckerman, four Vermont lawmakers drafted JRH 167. The resolution was a direct response to the controversial 1997 NOP proposed rules. The joint resolution aired a number of grievances of state lawmakers. First, the 1997 proposed rules created subpar and weaker standards than the NOFA-VT established in the previous decades. The permissible use of synthetic materials and the controversial big three (e.g., irradiation, sewage sludge, and GMOs) in production was not acceptable. Second, the NOP proposed rules that created a disadvantage for small organic farmers. The $400 fee and pesticide testing costs would financially destroy the long-term existence of small-scale organic farmers, a key growing group of Vermont (Vermont Senate Agriculture Committee 1998a; Vermont Senate Agriculture Committee 1998b). The state already saw increasing competition from large-scale growers. Looking at California's one-year certification process, it was easier for larger companies to meet. Vermont leaders believed that the national standards would eliminate the standard discrepancies and put smaller-growers at a disadvantage. Finally, lawmakers were alarmed by organic livestock rules. The proposed rules would permit livestock to be fed nonorganic feed and still be labeled as organic. The practice was not in accordance with most of Vermont growers and producers' organic beliefs.

The grievances aired by Vermont lawmakers were similar to the many 250,000 complaints lodged against the 1997 proposed rules (Vermont Senate Agriculture Committee 1998b). There was a truly deep concern that the proposed rules would hurt the organic market, a sector important to Vermont. Zuckerman operated one of the 170 organic farms in the state in late 1990s. This personal factor inspired him to write and sponsor the resolution. The state organic producers, predominantly small-scaled in nature, feared a decline in trust in the organic label and decline in annual results that could result from a diluted meaning. JRH 167 was a call to correct the federal standards after the state put forth its own approach to regulating the organic market in 1990, a producer-designed approach that was not incorporated into the federal approach.

The Demise of the Organic Advisory Board

At the beginning of the 2003–2004 legislative session, S.11 was introduced by Vermont senator Sara Branon Kittell. Kittell, a Democrat from Franklin County, sponsored and introduced the bill. S.11, also known as The Housekeeping Bill, included a number of language amendments in Title 6 of the Vermont statutes (Vermont Senate Agriculture Committee 2003). Title

6 V.S.A. § 181 was repealed, resulting in the termination of the Organic Farm Advisory Board on July 1, 2003.

The termination of the Organic Advisory Board was not a complicated or contentious issue among Vermont lawmakers. Termination of the advisory board was merely a routine maintenance procedure, as the name of the bill suggests. The implementation of the NOP final rule in 2002 preempted state organic regulatory policies. While many states decided not to terminate statutory language (see chapter 3), Vermont was one of nine states to dismantle existing law after NOP implementation. The particular cause behind the statutory repeal came down to federal supremacy. The existence of the Vermont Organic Advisory Board would be in direct violation of the NOP final rule. With a national law in place, there was no need or legal basis for maintaining an organic regulatory policy in the state (Vermont Senator 2015).

Since the repeal of the board, Vermont continues support for sustainable and organic agriculture research and education programs. In addition, policy makers have also pursued a number of other highly innovative sustainable agriculture and food labeling regulatory laws. In 2013, Vermont became the first state in the nation to adopt a non-contingency based GMO-labeling law (2014 Act 120). Other states in the region, including Maine and Connecticut, adopted laws requiring the identification of genetically engineered ingredients in foods before Vermont, but Vermont was the first state to require labeling without needing other states to adopt similar laws before implementation.

EXPLAINING ADOPTION

Organic regulatory policy in Vermont was a short-lived endeavor. Almost as soon as the Organic Advisory Board began meeting, it became clear that the federal law would eventually go into effect. The timing of initial enactment was in clear response to a rising tide of interest in organic goods, regional and vertical pressures, and a desire to distinguish Vermont's unique perspective on how best to regulate the market. Despite the advisory board never truly achieving its goal, Vermont lawmakers were quick to respond and pushback against federal regulatory efforts in 1998. Of course, Vermont respectively repealed its own organic regulatory laws once the NOP final rule went into effect. Many of the findings from chapter 3 are further elaborated upon in this case.

Internal Conditions

Political and economic conditions positively influenced the timing of Vermont's initial adoption decision and the decision to repeal the law in 2003.

The political environment was one that supported the efforts of NOFA-VT once the organic producers association was ready for government support. NOFA-VT was instrumental in pushing for legislation in 1990. S.277 was designed by organic producers for organic producers. The same farmer-on-farmer certification design seen earlier in California was the center of the Vermont bill because of NOFA-VT's involvement. NOFA-VT served a central facilitating role in determining regulatory efforts largely because the organization faced limited pressures for needing government involvement until the late 1980s. The organization came together regionally first before the market truly demanded certified "organic" goods in the mid-1980s. The producer-distributor relationship was truly built more on personal relationship and trust rather than an established, objective certification process (Former VOF Staff Member 2015; Gershuny 2016). Once the market began to grow rapidly, the grassroots organization and certification group finally embraced government involvement beyond financial and administrative support provided by Vermont's agriculture department.

The political conditions in the state aided in the bill development by serving as a necessary and sufficient role to NOFA-VT. First, the political culture is identified as predominantly moralistic with some elements of an individualistic culture (Elazar 1972 and 1994; Sharkansky 1969). This sort of political culture is similar to California and likely supported the organic movement in the region like that on the West Coast. While there was no university support, Vermont is likened to Northern California in terms of agricultural production interests and had a strong hippie-driven back-to-the-land movement, which resulted in strong development of a local, organic food market (Former VOF Staff Member 2015). Yet, one strong curiosity is that despite being a receptive culture to government involvement in the market, Vermont delayed action until 1990. This could be partially explained by the unique administrative partnership NOFA-VT established in the mid-1980s, a second political determinant.

Politicians and administrators gave overwhelming support to certification of operations and the adoption of a law that would protect producers. The Vermont Agency of Agriculture, Food, and Markets fully supported VOF and NOFA-VT as the groups established an excellent working relationship to regulating the market prior to 1990. The extension of grant, administrative, and marketing support from the agency to NOFA-VT and VOF helped expand the market in the state, and perhaps the region, by elevating the level of support for organic farming (Wonnacott 2015; Former VOF Staff Member 2015). In addition to agency support, organic producers found support among the ranks of Vermont lawmakers sympathetic to the organic farmer needs. Elizabeth Ready, a Democrat, helped constituents of her district secure legal backing for certification standards set forth by VOF. David Zuckerman, a

Progressive, is an organic farmer particularly concerned with the direction of the federal efforts when leading a resolution challenging the 1997 NOP's proposed rule. While both scenarios may be by chance, it is clear that left-leaning partisan support was a factor in pursuing organic policies and giving voice to Vermont's organic farmers at the national level.

External Conditions

Both horizontal and vertical dimensions of diffusion were present in the case of Vermont. Each bill and resolution had a clear mixture of regional and national pressures. Yet, federal pressures appear as a more immediate concern for initial adoption, repeal, and general dissent. Regional and national interaction patterns among states did factor into Vermont's legislative design.

Horizontal Diffusion

With NOFA-VT serving as a key driver for Vermont's organic policy adoption, it is this organization that sheds light on how other states influenced Vermont's legislative decisions. Prior to the mid-1980s Vermont's organic farmers operated within a tight-knit Northeastern community. New Hampshire farmers were by far the closest as they were co-founders of NOFA in 1971. Other states were also significant, such as New York and Massachusetts, by serving as hubs for organic produce sales (Gershuny 2016). By 1990, NOFA's chaptered membership base extended to all Northeastern states except Maine, who is cited as doing their "own thing" (Former NOFA-VT Staff Member 2015). The regionally based relationships ranged over time from friendly to somewhat hostile. The breakdown of NOFA into different chapters in 1979 signifies differing opinions and power struggles within the group, but collectively NOFA has cooperatively operated as a chapter-based organization since.

Vermont's connection with other states outside of the region started in the mid-1980s. As one individual described it, the channels of communication became wide open at this time with Vermont organic leaders coming in contact or learning about the regulatory efforts in several other states. Before the mid-1980s, the communication between various regions was "spotty" at best (Former NOFA-VT Staff Member 2015). Eleven states were directly referenced either in historical documents or interviews as examples of other regulatory and certification efforts including California, Colorado, Iowa, Maine, Massachusetts, Minnesota, New Hampshire, Oregon, Texas, Washington, and Wisconsin (Former NOFA-VT Staff Member 2015; Former VOF Staff Member 2015; Gershuny 1990; Stevens 1990). References varied from discussion on leading innovators in organic production and certification, the state-operation of certification programs, and policy content design, such as designating the implementing state agency or general rule development on such matters as transitionary periods.

Considering the close relationships established and general awareness of other state efforts, it is somewhat surprising Vermont government action lagged behind so many other states. Internal conditions were receptive and ideal for an early innovator, but the particular regulatory design enacted may be reflective of the underlying cause for delay. The emphasis on a producer-centered law with a cooperative relationship between government and certifiers was crucial for lawmakers and organizers. It is a stance that was a dissenting minority opinion nationally among states. So despite heightened knowledge and interaction with others, Vermont's bill design likely minimized regional and national interaction diffusion pressures and elevated the tension between state and federal efforts.

Vertical Diffusion

The federal government played a strong and significant role in Vermont's organic policy story. Kathleen Merrigan's Vermont visit in 1989 shook the local organic community into action. Growers supported the idea of a national law but did not support the proposed direction Merrigan discussed (Wonnacott 1989). The federal law was too consumer focused and removed the power of producers to determine what organic meant. The federal government would set a minimum standard and chose to dismiss regional and state variation in production. In turn, this limited the ability of growers and producers to truly be more or less organic (Former VOF Staff Member 2015; NOFA-VT Staff Member 2015; Wonnacott 1989). Vermont's dissenting opinion led to a state policy design that was in response to federal efforts, thereby suggesting a positive relationship between state and federal adoption decisions.

JRH 167 and the repeal of the Vermont Organic Standards Board were also a response to federal efforts. Lawmakers were deeply concerned about the direction of federal law after the 1997 proposed rules. The resolution outlined a number of grievances that echoed other public comments submitted. While resolutions were not included in the statistical analysis, Vermont's resolution signifies overarching concerns among state lawmakers about the quality and direction of federal law. Once the NOP was implemented, state lawmakers responded once again to federal activity. The implementation of the NOP in 2002 has a positive relationship with Vermont amending, or rather terminating, state policy in 2003.

The interaction between Vermont's organic policy action and federal activity is significant to explaining why the state pursued regulations. Federal development of a law spurred state action, although uncertainty did surround the enactment of the OFPA. Once adopted, federal rule making and final implementation acted as key determinants in explaining state adoption behavior. This case study confirms the significant role of vertical diffusion determinants likely played for many middle-of-the-pack adopters and later reinvention or amendments of state policies.

Issue Salience

The attention to organic food and agriculture was elevated and growing by the time S.277 was adopted. Like California, Vermont organic producers were concerned with increasing fraud in the marketplace as the public became enlightened to the dangers of conventional agriculture and benefits of organic agriculture. Focusing events like the carrot caper scandal, Meryl Streep's *60 Minutes* appearance, and the case of George Crane's fraud were significant to members of NOFA-VT and VOF. Yet, these cases were only marginally significant compared to other factors. In essence, the role of issue salience was not widely cited as a primary determinant but rather a peripheral factor that contributed to a broader systemic awareness that accelerated state organic policy adoptions nationally and regionally.

CONCLUDING VERMONT

Vermont adopted its organic regulatory legislation at the height of the critical mass of diffusion. Increasing attention to the organic agriculture, broader networking among organic growers, and congressional development of national standards pushed lawmakers in Vermont to act. NOFA-VT and its certification branch, VOF, were the primary policy actors driving the initial adoption by acting a medium to filter activities occurring locally and nationally. The state political environment fostered a healthy environment for NOFA-VT to work collaboratively with the state's agriculture department prior to seeking a legislative bill. The lawmakers and administrators saw a clear benefit in fostering and protecting a burgeoning agricultural market. In turn, S.277 was reflective of the standards and value of organic producers in the state. Once S.277 was enacted, lawmakers, some motivated by personal connections to organic farming, were quick to respond and react to federal rule-making activities, which were at odds with how state foresaw organic standards. The story of Vermont's organic regulatory legislation details a state that deliberately waited to enact government-backed standards and sought to respond to federal organic policy efforts until enactment of the national program.

NOTES

1. The Progressive Party is described as keeping the Democrats liberal. Republicans in the state are described as being socially tolerant but fiscally conservative (Former VOF Staff Member 2015).

2. According to the most recent 2012 Ag Census, there is 1.25 million acres in farms with cropland now constituting 39 percent of total farm acreage.

3. For a more in-depth discussion of Vermont's biophysical regional differences see Elizabeth Thompson's (2002) "Vermont's Natural Heritage: Conserving Biodiversity in the Green Mountain State." https://www.uvm.edu/rsenr/sal/vbp/VBP.pdf.

4. The 2008 economic crisis is a notable exception to stability. Following the economic decline, dairy farmers faced historic lows in pricing.

5. NOFA once stood for the Natural Organic Farmers Association. The change in name reflected regional focus and the switch from farmers to farming broadened the membership base to incorporate gardeners and non-farmers interested in organic practices (Wonnacott 2015).

6. For a historical overview of the split, see Gershuny's *Reclaiming the "O" Word.*

7. The OCIA was a producer-focused set of standards. Comparatively, the OCIA standards were estimated to be 97–98 percent similar to standards developed by CCOF (NOFA-VT Staff Member 2015).

8. Commissioner Albee and Steve Justis are credited as providing significant support for the partnership between the Agency of Agriculture, NOFA-VT, and VOF (Stevens 1990).

Chapter Six

Organic in Georgia

A Laggard

The history of organic legislation in Georgia is briefer and not as well documented in comparison to California and Vermont. Georgia's first organic food and agriculture bill was adopted in 2000, twenty-one years after California's initial legislative adoption and ten years after Vermont's first bill. Beyond the comparative differences, the case of Georgia is intended to shed some light on what conditions influence laggard adopters and the adopters in the Southeast region. From a national perspective, Georgia can be considered a lagging adopter, who only pursued legislation ten years after the critical mass ended. Thirty-one other states had existing state organic food and agriculture statutes prior to 2000. Compared to other states in the Southeast region, Georgia's initial adoption lagged behind Florida, Kentucky, and North Carolina. Mississippi adopted organic legislation for the first time in 2000 with timing identical to Georgia. Since 2000, no other state in the Southeast has adopted organic legislation.

The case of Georgia provides depth for understanding why some states and the Southeastern region did not eagerly pursue organic regulatory policy. The statistical results in chapter 3 do not provide much guidance for why certain states waited to pursue organic regulatory policies until after a critical mass was achieved and OFPA was adopted. Evidence from this case sheds some light on how delayed federal implementation inspired some states to adopt organic regulatory legislation, the significant role of university research and presence of organic networks in successful enactment of state policy, and the challenges associated with regional perceptions of organic foods.

This chapter begins with a brief overview of Georgia that provides a framework for understanding the conditions under which policy decisions are made. Second, the historical development of the Georgia agricultural industry is evaluated with an emphasis on the development of organic agriculture in the state. Next, an overview of Georgia's three organic legislative bills is provided. Finally, the chapter finishes with an evaluation of significant factors of adoption in Georgia and concluding remarks. The results indicate that federal policy and consumer and producer demand, perhaps driven by economic and political factors, influenced Georgia's initial adoption of legislation in 2000. Amending legislation, however, is influenced by federal implementation and broader financial considerations to maintain financial sustainability of the program.

GEORGIA POLITICAL AND SOCIO-ECONOMIC CONDITIONS

Georgia, also know as the Peach State, is similar to many other states in the Southeastern United States. The political and economic environment are markedly conservative, pro-business, and hierarchical. Hubs of liberal support do exist within the state, but, like other states in the region, Georgia remains part of the larger southern voting bloc Republicans can generally depend on each election cycle. Similar to California and Vermont, agriculture has a significant history in the state but is not a major contributor to the overall economy. However, unlike California and Vermont, grassroots interest in organic farming is historically absent in the state with it remaining a fringe practice.

Politics in Georgia

As part of the Solid South,[1] Georgia largely identifies as a Republican state with hubs of Democratic support existing (Georgia Secretary of State, n.d.; Lang and Pearson-Merkowitz 2013). A cluster of Southwestern counties and the metropolitan areas surrounding Atlanta, Athens, Augusta, and Savannah have consistently voted majority Democrat since the 2000 presidential election (see appendix E). Only minor shifts in partisanship voting occurred since 2000; the 2004 voting results show fewer counties in Georgia voting majority Democrat than any other presidential election year. Except for counties in the southwestern portion of the state, Democratic-voting areas have higher per capita income and higher population density (see appendix E).

Compared to other states within the region, Georgia's in-state political environment is on par with regional averages and norms (see appendix E). Georgia's political culture is identified as traditionalist with a moderately strong individualistic strain, which correlates to an 8.80 on Sharkansky's political culture scale (Elazar 1994, 284; Sharkansky 1969). Regionally, the average political culture is 8.415 indicating most states are predominantly traditionalist. This means that Georgia's political culture is dominated by a hierarchal system that maintains strong social order and familial ties. Often, certain families and elites have more political power. However, the strong individualistic strain indicates a cynical attitude toward government where the protection of individual rights is sought.

Georgia's state legislature professionalism conforms to the regional ebb and flow over time. However, the actual state ranking, which was .135 in 2010, is low compared to other states including Florida and North Carolina. In general, Georgia's state legislature is less professionalized than other states in the nation and within the region. Finally, state government control is also similar to other states within the region. Georgia has had Democratic, Republican, and split-control of state government from 1990–2010. Like Florida, Georgia state government transitioned from Democratic control to Republican control over time. In sum, the political conditions in Georgia are comparable to other states within the region but different from other regions in the United States. Georgia is inherently more traditional, less professionalized, and leans toward Republican political candidates.

Agriculture in Georgia

The history of agriculture in Georgia is identified as complex and defined by a series of terms including "soil-abundance, fertility, misuse, abuse, and ultimately, conservation and regeneration" (University of Georgia Libraries, n.d.). The state has a long history of agricultural production starting before European settlement by General James E. Ogelthorpe and the English colonists (Flatt 2013). Prior to European settlement, the Cherokee and Creek Indian tribes farmed corn, beans, melons, and fruits (University of Georgia Libraries, n.d.). They also collected nuts from pecan, walnut, and hickory trees that grew wild. Early European settlers in the state, ranging from the English, Germans, Portuguese, and Scots cultivated a variety of vegetables and indigo in a thirty-mile wide area near Savannah (Bonner 1964, 3; University of Georgia Libraries, n.d.).[2] Most of the early settlers lacked agricultural skills and the use of indentured servants was common practice until the introduction of slavery after 1750 (Bonner 1964, 3; University of Georgia Libraries, n.d.).[3]

After the slavery ban repeal, a host of new crops would be grown as the new labor force became available. Cotton, along with tobacco and sugar cane, would be the most significant crops after the Revolutionary War (Bonner 1964, 49; University of Georgia Libraries, n.d.). Mechanical, laborsaving machines only helped to spur the cotton industry's growth (Bonner 1964, 96–105). However, the dependency on cotton as a staple crop would eventually lead to devastating financial and economic consequences. Farmers, particularly those in Georgia's cotton belt, would be affected.[4] Poor agricultural practices, supported by political and social infrastructure, led to soil degradation starting in the mid-1800s. Clear-cutting forests for timber and lack of crop rotation led to topsoil erosion and soil quality decline (Bonner 1964, 63; University of Georgia Libraries, n.d.). Poor soil quality led to a decline in crop quality and decreased crop yields. Combined with the overproduction of cotton nationally, Georgia's cotton farmers faced declining prices for the cotton crops yielded under poor soil conditions.

Several efforts over the years attempted to diversify Georgia's crops and improve agricultural production in the nineteenth and twentieth centuries but cotton remained king (Bonner 1964, 38; Flatt 2013). The final demise of cotton as a staple crop is attributed to the boll weevil and World War I (Flatt 2013).[5] Low cotton prices post-World War I and the spread of the boll weevil into Georgia led to drastic decreases in production that would continue through the Great Depression. Diversification of Georgia's agriculture would begin during World War II with the support of the Agricultural Adjustment Act of 1933 and the Soil Conservation and Domestic Allotment Act of 1936 (University of Georgia Libraries, n.d.). Both bills provided financial compensation for farmers to grow soil-conserving crops and to engage in practices that would rebuild soil quality.[6] The Georgia State Legislature also provided help through the establishment of Soil Conservation Districts in 1937 (Brown 2002b, 79–82). District activities included draining swamps, managing forests, promoting soil conservation, and improving pastures. Civic organizations, the media, and churches helped to carry the message of conservation and soil stewardship.

Since the restoration of soil quality in Georgia, the agricultural portfolio of the state has become more diversified. Although cotton remains a top-earning commodity in the state, peanuts, pecans, apples, peaches, blueberries, and soybeans are some of the most important commodity crops for the state as well (University of Georgia Cooperative Extension 2012). Georgia is the top producer nationally for peanuts, chickens, pecans, and watermelons (Georgia Farm Bureau 2014). The agricultural industry, including forestry, employs one in six Georgia workers (Georgia Farm Bureau 2014) but contributes marginally to the states overall GDP. In a national context, Georgia ranks in the top ten states for agricultural productivity and the top twenty for farm outputs in 2011 (ERS 2013a, 2013b). Furthermore, the southwest region of Georgia contains one of the more predominant areas for fruit and vegetable production in the country (National Agricultural Statistics Service [NASS] 2012a).

Agricultural regions in Georgia can be designated in several ways based on geographical features, elevation, and soil type. Figure 6.1 demonstrates one approach to mapping agricultural regions within the state. The map represents

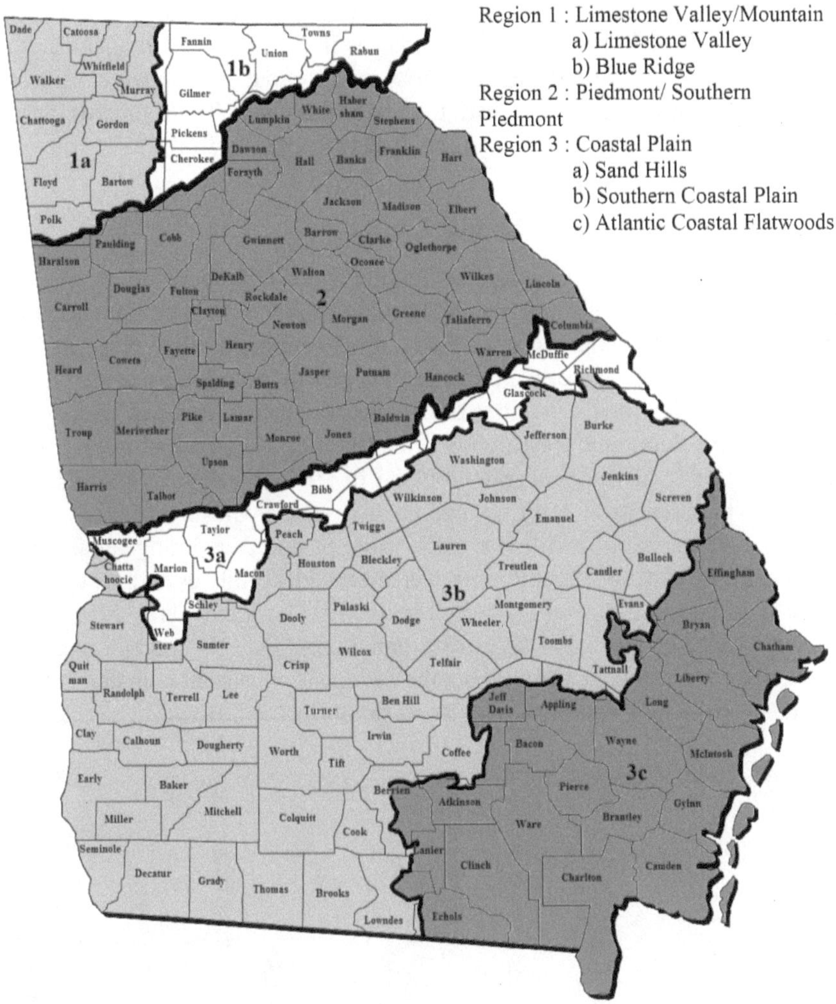

Figure 6.1. Agricultural Regions of Georgia
Created by the author from data compiled from Hancock et al. (2011).

three main geographic regions and six soil provinces in the state used by crop and soil scientists at the University of Georgia (Hancock et al. 2011). The three main geographic regions represent climatic differences in the state such as higher elevations in the limestone valley and lower elevations in the coastal plain. The soil provinces represent differences in organic matter content, texture, and drainage. The Piedmont region is where severe soil erosion occurred as a result of row cropping, including cotton and tobacco, in the 1800s and early 1900s. The Southern Coastal Plains, which includes the southwestern area comparable to the cotton belt, contains heavier and more fertile soils. Even accounting for the relatively diverse differences in regions and soil qualities, the number of farms is fairly balanced across the state but with a greater share of land in farms located in the southern portion of the state in the Southern Coastal Plain and Atlantic Coastal Flatland provinces (NASS 2012b).

Organic Agriculture in Georgia

The history of organic agriculture in Georgia is not well documented. There is no precise indication of when organic farming began in the state even accounting for soil-quality depletion at the turn of the twentieth century. Nevertheless, since 1997, the Georgia organic market has grown but remains ranked in the bottom ten states for the number of certified organic operations.[7] According to the USDA Economic Research Service (2013c), Georgia had one certified grower and 572 acres of organic cropland in 1997 (see figures 6.2 and 6.3). By 2011, the number of certified organic growers grew to seventy-five with a total of 5,483 acres of certified organic crop, pastures, and rangeland. While the number of certified organic acres has increased steadily over time, the peak number of certified growers occurred in 2007 when there were 121 certified growers representing 2,544 acres of organic cropland. No specific reason is identified for the decline in 2007, but the decline is consistent with the national trend that coincides with the Great Recession (Greene 2013).

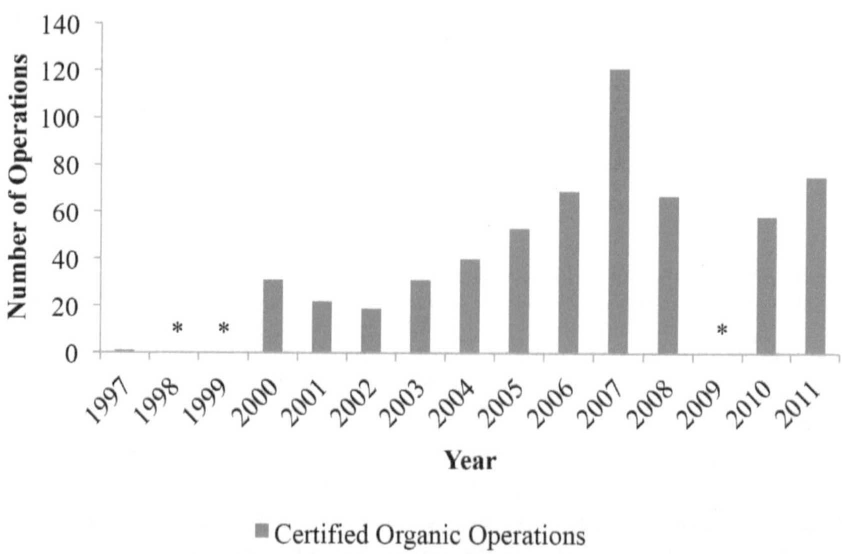

Figure 6.2. Certified Organic Operators in Georgia
Created by the author with data compiled from ERS 2013c.

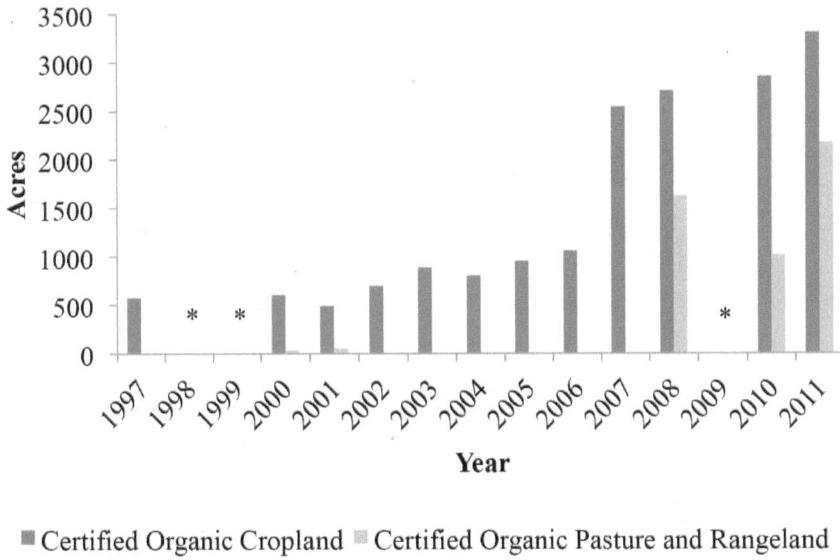

Figure 6.3. Certified Organic Crop, Pasture, and Rangeland in Georgia
Created by the author with data compiled from ERS 2013c.

The distribution of certified organic operators is not even across the state. A majority of certified operations are located either in northern Georgia, surrounding the greater Atlanta metro area, or the south central portion of the state near Savannah, where consumer interest is identified as more intense (Georgia Department of Agriculture [GDA] Staff Member 2013; NASS 2013). Large-scale organic operations, producing agronomic crops such as feed grains, peanuts, and hay, are located in southern Georgia (Agriculture Specialist 1 2013). An overwhelming majority of certified organic operations are crop growers and handlers. Out of the 166 certified operations, there are eighty-three certified operations as crop growers, eighty-two certified handling operations, and four are livestock. A few of the businesses are certified in two categories. Figure 6.4 demonstrates the distribution of certified operations in the state as of January 2013.

Two organizations within the state, Georgia Organics and the Georgia Crop Improvement Association, are identified as significant to organic production in the state (Former GDA Staff Member 2 2013; Agriculture Specialist 1 2013). Several interview participants deferred answering specific questions and recommended contacting individuals from each organization.[8] However, limited information exists or is provided detailing the role of Georgia Organic and the Georgia Crop Improvement Association in organic farming in the state. Neither organization was involved in promoting or

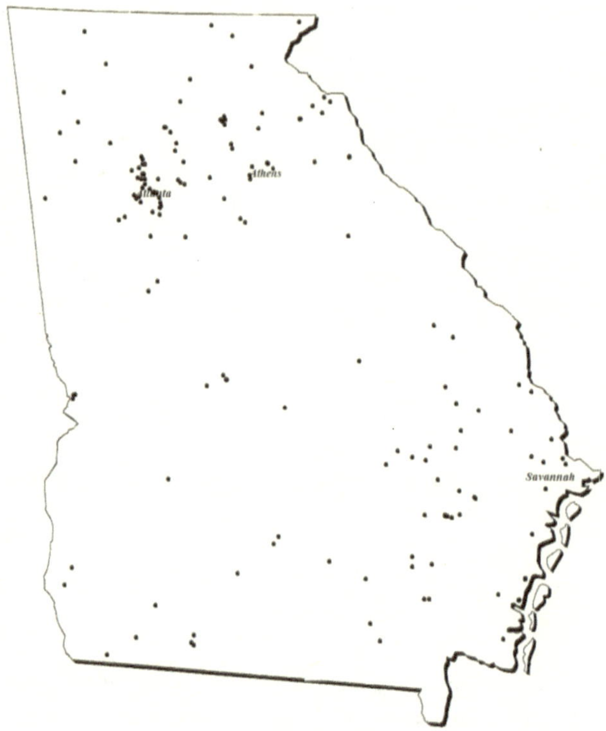

• = Approximate location of certified organic operations

Figure 6.4. Certified Organic Operations in Georgia
Created by the author based on information from NOP 2013b.

developing the initial organic food and agriculture bill adopted in the state in 2000, but both organizations are involved with the promotion of organic production since 2002. Emerging out of a 1970s growers' association, Georgia Organics is a non-profit organization seeking to create sustainable local food systems in Georgia (Georgia Organics 2012). The organization plays a role in the state's outreach as well as promotion of its organic program and certification option after the official implementation of the NOP (Former GDA Staff Member 2 2013).

Conversely, the Georgia Crop Improvement Association also helps with outreach and promotion of the state's organic efforts albeit in a different manner. The association is a farmer organization created in 1946 (Georgia Crop Improvement Association Organic Certification Program 2013). Two bills, 1956 HB 104 and 1997 SB 583, made the association the legal certifying agency in Georgia. The intent of the association is to promote more efficient agricultural production through the use of better quality seeds. Organic certification is one of several certification programs the association offers.[9] Of the twelve organic certifiers operating within the state, the Georgia Crop Improvement Association certifies nearly half of the certified operations in the state as of January 2013 (NOP 2013b). Other major certifiers in the state include CCOF, Clemson University Department of Plant Industry, Quality Assurance International, and Quality Certification Services.

Georgia can be considered a latecomer to expressing interest in organic food and agriculture compared to other states in the country. Several interview participants indicated that organic was a term that was not always salient in the state (Agriculture Specialist 1 2013; Former Organic Food Advocate 2013). Evidence gathered in Georgia indicates that the support for organic products within the state has only come after a few decades of normalizing the term. Organic was not a common term used in the South for much of the past four decades (Agriculture Specialist 1 2013; Former Organic Food Advocate 2013; GDA Staff Member 2013).

From a consumer perspective, organic food has been met with skepticism and could even be referenced as a "California thing," indicating organic as an outsider product that isn't acceptable to the local culture (GDA Staff Member 2013). Indeed a consumer preference survey conducted in 1989, the same year of the carrot caper scandal in California (see chapter 4 for more detail), demonstrates that 61 percent of the surveyed population expressed preference to buy organic (Misra, Huang, and Ott 1991).[10] However, appearance of fresh produce was more important than being organic; over half of the respondents expressing organic preference would not buy "sensory defect" produce. Furthermore, most respondents did not provide a precise dollar amount for willingness to pay. The survey suggests that some Georgia consumers, namely older and wealthier white females, expressed a preference for organic products but may not be willing to pay or overlook cosmetic deficiencies of organic goods. This same preference may still exist today. When asked about the comparative perception of local versus organic food, a sustainable food advocate (2013) stated, "They are (more open to the idea of local compared to organic). That's because there's stigma attached to organic." In essence, consumer preferences in Georgia, and even throughout the Southeast, are slow to buy into the organic market, but some hubs of organic activity do exist within the state. Atlanta, Savannah, and Athens, home to the University of Georgia, are where communities have been more likely to participate in the organic market (Agriculture Specialist 1 2013; Sustainable Food Advocate 2013).

From a farming perspective, the increase in consumer demand nationally, since the late 1990s, mimics the increase of certified organic farms and

Figure 6.5. U.S. Annual Organic Sales
Created by the author based on information from Organic Trade Association (2011; 2016).

operations in the state. Figure 6.5 demonstrates the growth in the U.S. market by annual consumer sales. While correlating annual state consumer sales is unavailable, the U.S. Ag Census tracked state level organic farm sales in 2002 and 2007. Reported annual sales for Georgia increased from $671,000 in 2002 to $2,042,000 in 2007 representing a 204 percent increase (NASS 2012a). The growth in consumer demand at the national level, based on data from the Organic Trade Association, is 136 percent growth from 2002 to 2007; whereas the U.S. Ag Census data on organic farm sales indicate a 335 percent increase in organic farm sales from 2002 to 2007. The growth in the Georgia organic market coincides with a larger national trend. However, as the national market grew, Georgia's organic market nearly doubled during the same time period. Both trends demonstrate increasing interest in the organic market, but Georgia's dramatic increase suggests a rapid interest.

As the growth in the organic market occurred, so did the number of farms in Georgia However, despite an increase in the market, support for organic farming within the state is not strong, and farmers may be less willing to engage in organic practices. The dominance of large agribusinesses extends to GDA and even the research and teaching objectives at the University of Georgia (Agriculture Specialist 2 2013; Former Organic Food Advocate 2013). Funding and support for agricultural research that is predominantly geared toward conventional practices may stifle interest in organic production.[11] Furthermore, farmers may hesitate to transition land out of production for three years to meet organic requirements, considered a niche market to some farmers (Former GDA Staff Member 1 2013; Agriculture Specialist 1 2013). In general, growing conditions within the state can create difficulties to farm organically, and the economic consequences for attempting transition may not be worth the risk. For organic farmers that already existed within the state, Georgia's bill was an effort to protect the market and ensure consumer confidence in the organic label and the third-party certification scheme.

A HISTORY OF ORGANIC FOOD AND AGRICULTURE LEGISLATION IN GEORGIA

Since 2000, Georgia has adopted four bills pertaining to organic food and agriculture. Only three of those bills substantively changed Georgia's state statutes on organic food and agriculture.[12] Initial legislation was adopted in 2000 (see table 6.1). Substantive modifications to existing state organic food and agriculture statues occurred with the adoption of amending legislation in 2002 and 2010. Both amending adoptions resulted in minor changes to the structure of the statutes. To detail the development of Georgia's organic legislation, this section will first cover the 2000 Organic Certification and Labeling Act. Next, the details of the modification legislation are considered including related legislation and administrative rules after the 2010 amending adoption.

Table 6.1. Georgia Organic Food and Agriculture Legislation

Year	*Bill*	*Sponsored/Introduced**
2000	SB 477 *Organic Certification & Labeling Act*	Harold Ragan [D] & Michael Meyer von Bremen [D]
2002	SB 361	Harold Ragan [D] , Faye Smith [D], Donzella James [D], Seabaugh [R], and Peg Blitch [D]
2010	HB 1055	Kevin Levitas [R], Tom Rice [R], Matt Ramsey [R], Jim Cole [R], and Richard Smith [R]

**Political Party – [D] =Democrat; [R]= Republican*

Source: Data compiled by author.

2000 Organic Certification and Labeling Act (SB 477)

In February of 2000, SB 477 was introduced and read in Georgia's State Senate for the first time. The bill, also known as the Georgia Organic Certification and Labeling Act, amended Title 2 of the Official Code of Georgia to include regulatory language for the identification and certification of organic food and agriculture inputs and outputs during the production, distribution, and processing stages. Table 6.2 summarizes major aspects of the bill. The

Table 6.2. Summary of SB 477/ OCGA § 2-21

Aspect	*Summary*
Define Key Terms & Labeling Limitations	Includes definitions for certification, feed, and organic among other referenced terms (OCGA § 2-21-2). Identifies parameters for organic labeling (OCGA § 2-21-4).
Define Toxicity Levels	Limits organically labeled products to contain no more than 5 percent of substances considered toxic for the FDA, EPA, or USDA (OCGA § 2-21-3).
Record Requirements & Inspection	Requires producers, brokers, distributors, and processors to keep detailed records of products labeled as organic (OCGA § 2-21-3). Subjects any product to inspection using state or federal regulatory requirements (OCGA § 2-21-4).
Certification	Requires certification for organically labeled goods including food, feed ingredient, article, commodity, or product (OCGA § 2-21-4).
Rule Development & Enforcement	The Georgia Department of Agriculture is responsible for enforcement of the bill and developing any reasonable standards as necessary. GDA establishes fees to be collected for implementation (OCGA § 2-21-6). Violation of the organic law will be guilty of a misdemeanor (OCGA § 2-21-8).

Source: Data compiled by author from state bill and statute information.

final version of the bill designated the Georgia Department of Agriculture (GDA) as the enforcing agency and responsible for promulgating rules for implementation. Definitions, labeling requirements, and maximum levels of toxic material were established to provide clear guidelines. All certified entities must obtain certification from GDA or a GDA approved entity. Furthermore, all certified entities must maintain detailed records of their organic production and handling practices and processes. To implement the law, GDA was authorized to establish fees that would support the services provided through implementation.

Harold Ragan and Michael Meyer von Bremen co-sponsored the bill. Both state senators were Democrats representing districts in the south-southwestern area of the state and members of the Senate Agriculture Committee. GDA staff members approached Senator Ragan, chairman of the Senate Agriculture Committee, to sponsor the bill. The purpose of SB 477 derived from concerns among Georgia's organic farmers and the GDA about establishing laws in the state to promote and protect organic products. As two former staff member stated, the GDA wanted to regulate the organic industry to make sure labeling was accurate and in accordance with federal requirements (Former GDA Staff Member 1 2013; Former GDA Staff Member 2 2013). No intention was made to go above and beyond the federal requirements (Agriculture Specialist 2 2013). The bill was a truth in labeling effort spurred by demand from the public and certain producers interested in pursuing organic certification. Concern was raised that third-party certifiers might not be conducting truly independent evaluations and audits.

SB 477 was amended once during the 2000 legislative session to include explicit language regarding the regulatory nature of the bill and to remove language describing organic food and feed rules. As introduced, SB 477 created OCGA § 2-21-3 to include specific details of what constitutes organic food and feed. As one former organic advocate indicated (2013), the state was on the forefront of pushing for federal regulation of organic food and feed but this language was removed in the final version of the bill. After amendments, SB 477 passed unanimously in both the Georgia House of Representative and Senate. No state legislator opposed passage of SB 477.

Reinventing Georgia Organic Law: Amending Legislation

Georgia modified its organic law twice since the initial adoption in 2000. Both bills included provisions that modified OCGA § 2-21-4. Adopted in 2002, SB 361 replaced the entire subsection of the state organic state statutes. Table 6.3 outlines the key amendment provision. SB 361 removed certification seal requirements for food and feed ingredients but maintained requirements for food and feed to comply with other organic regulations and certification requirements. The bill also added additional requirements for any individual involved in the production, processing, and distribution of handling of organic products. Each organic entity would be required to register annually with GDA starting in 2003. Fees collected from registration would go toward implementation of the law. SB 361 passed unanimously.

HB 1055, adopted in 2010, would increase the minimum and maximum fee structure from $25.00/$500.00 to $75.00/$1000.00. The amendment was part of a larger bill focusing on a variety of amendments bundled as the "Georgia Taxpayer Relief Act of 2010" (Weber 2010). The House Committee on Ways and Means added the amendment to the organic fee structure. HB 1055

Table 6.3. Summary of SB 361/ Revision of OCGA § 2-21-4

Aspect	*Summary*
Certification Seal	Removes requirement for food and feed ingredients to bear the official seal of the certifying entity (OCGA § 2-21-4(d)).
Registration	Requires any individual producing, processing, distributing, or handling organic products to register with the GDA on or after January 1, 2003. Establishes annual registration fees no less than $25.00 but no more than $500.00 (OCGA § 2-21-4[e]).

Source: Data compiled by author from state bill and statute information.

passed with some contention related to other non-agricultural measures in the bill such as the elimination of the retirement income tax and property tax. The adjusted organic fee structure was not controversial.

EXPLAINING ADOPTION

The story of Georgia's organic food and agriculture legislation is a short and relatively uncomplicated story. Perhaps the most interesting fact to note is that most individuals contacted for interviews did not know the state of Georgia enacted a law regulating organic food and agriculture. Of the four interview participants that did know of organic legislation, each was intimately involved in the adoption or implementation of the 2000 bill. However, even accounting for the lack of awareness or documented procession of state legislation, the case of Georgia still highlights actions by organic diffusion laggards.

All three legislative adoptions in Georgia were influenced by both internal and external conditions. The initial adoption in Georgia was influenced by federal activity, consumer and producer demand, and perhaps shaped by partisanship and political culture. Furthermore, some evidence also suggests regional influence through an organic industry network, which is considerably smaller and less active compared to the cases of California and Vermont. The effects of regionalism may also play a factor in why increased agricultural productivity is significant, as well. Soil degradation followed by attempts to rebuild soil quality in the twentieth century in the Southeast may have aided the increased agricultural productivity in the region. For Georgia, the soil rejuvenation helped to maintain the importance of agriculture in the Georgia economy.

For amending adoptions, federal implementation was the primary cause for the 2002 adoption, a finding consistent with the other cases. The Georgia response was not to terminate existing statutory language and laws but to accommodate federal supremacy. The 2010 adoption was not clearly linked to any consideration except maintaining financial sustainability of the program, perhaps the result of post-Great Recession financial consequences and Republican-controlled state government.

Internal Conditions

Both internal economic and political condition contributed to adoption decisions in Georgia. For the 2010 amending adoption, no clear indication was made for why fee structures were changed. While it could be the result of partisan politics, it would be overly presumptuous to assume that the partisan debate surrounding the taxpayer relief bill was centered on raising fees for organic growers and handlers. The need for tax reform may signify larger systemic factors at play. The Great Recession is a factor that may have led the Republican-controlled state government to pass a bill to provide more financial sustainability within the state. While not directly assessed in this analysis, the effect of overall state economic health may have contributed to adjustments to the organic registration fee structures to help minimize the cost of the state to implement organic food and agriculture policy. Other fee structures included within the bill also accounted for the need to generate revenue to offset administrative costs.

For the initial adoption of SB 477, indirect consumer and producer demand drove the desire for the GDA to seek regulations to ensure truth in labeling claims (Former GDA Staff Member 1 2013; Former GDA Staff Member 2 2013). Consumer demand and production of organic food, particularly from metropolitan hubs, is likely what was under consideration. This finding confirms the expectation of Winders and Scott that urban consumers have the ability to impact agricultural policy direction (Winders and Scott 2009, 80). It may be the increasing number of organic farms correlates with increasing consumer interest. Furthermore, farms that did transition into organic did so because the farm type and other relevant conditions (e.g., labor intensity level) may have led to an easier transition to organic farming practices.

The metropolitan hubs shed some light on additional conditions that likely led to adoption. In particular, wealth and political condition may have favorably led to the pursuit of regulations. Moreover, the role of the state's local food program, Georgia Grown, may have also created a gateway for perked consumer interest within key Georgia metropolitan areas. Wealth and political culture may be directly linked to rising consumer concern. The Atlanta, Athens, and Savannah areas are wealthier, with median household incomes ranging from $35,000 to over $75,000, and have political cultures that are more progressive compared to other parts of the state (U.S. Census Bureau 2013). On average, counties surrounding Atlanta, Athens, and Savannah have higher median household incomes, almost 1.5 to 2 times higher than rural areas, and have voted Democratic in presidential election years since 2000. Higher incomes not only correlate with higher education levels but also better health outcomes (National Center for Health Statistics 2012). Collectively, the factor of wealth in Georgia could explain the influence of consumer demands in Georgia metropolitan areas. Organic product consumers are usually wealthier, more educated, and concerned, in part, with health concerns. Therefore, organic consumer demand is situated in or near Atlanta, Savannah, or Athens and consumer demand is primarily driven from these areas.

The political activity of the Atlanta, Savannah, and Athens metropolitan areas can also help to explain why these regions are significant to the adoption of SB 477. As a whole, the state is identified as having a political culture that is predominantly traditionalist with a moderately strong individualistic strain (Elazar 1994, 284; Sharkansky 1969). Both Democrats and Republicans have controlled the state government. In 2000, Democrats controlled the state despite a majority of the state voting Republican in the presidential election. Democratic voting patterns, however, are consistent in several metropolitan areas, which are also wealthier. The areas surrounding Atlanta, Savannah, and Athens have higher per capita incomes and are more likely to vote for Democrats, whose members in the Georgia state legislature sponsored the 2000 and 2002 organic bills, than rural areas of the state. The combination of wealth and local political acceptance of innovative policies may contribute to a rise of interest in organic, but a third explanation still exist—the Georgia Grown program.

The Georgia Grown program is a marketing program operated by the GDA. Its purpose is to promote locally grown, but more specifically, Georgia-grown products. Likewise, considerable efforts have been made to improve accessibility and to promote farmers' markets in metropolitan areas such as Atlanta. Improving the quality and access to farmers' markets may result in healthier food purchases and also promotes local farmers. Promotion of local farmers may also help to explain a correlation between adoption and an increase in agricultural productivity within the Southeast region and in the state. The emphasis with both programs is locally based food, an idea more popular in the Southeast (Former GDA Staff Member 1 2013; Former Organic Food Advocate 2013; GDA Staff Member 2013; Sustainable Food Advocate 2013). As several participants indicated, the connection between local and organic is crucial to sustainable agriculture, but the local component may also serve as a gateway for consumers to become interested in organic (Former Organic Food Advocate 2013; Sustainable Food Advocate 2013). Specifically, the local food movement in Georgia contributes to consumers questioning how their food is made. This in turn may lead to consumers becoming more open and interested in the idea of organic food purchases. Organic has openly been indicted as a "California thing" or non-southern concept. Indeed, southerners may be generally suspicious of organic food (GDA Staff Member 2013). Bridging connections to organic via locally grown goods can de-stigmatize negative perceptions. In sum, the adoption of SB 477 in Georgia is the result of both political and economic pressures within the state and the consideration of federal activity in the policy area.

External Conditions

A combination of both horizontal and vertical dimensions of diffusion also contributes to adoption decisions. The influence of the federal government was confirmed as directly impacting the 2000 and 2002 bills. Less clear is the effect of other states on Georgia lawmakers. There is evidence to suggest that regional pressures may have contributed to diffusion within the region. However, this evidence cannot definitively be confirmed.

Horizontal Diffusion

Two states, North Carolina and Florida, are mentioned as influential to Georgia's agricultural policy (Agricultural Specialist 2 2013). Specifically, Georgia, Florida, and North Carolina shared similar behavioral patterns in adopting certain agricultural practices (Agriculture Specialist 2 2013). Both states mirrored the movement in and out of cotton production, the move into peanut production, and the move into organic production. The organic production network within the Southeast may be similar, as well. While not clearly linked, two interview participants noted the significance of the Carolina Farm Stewardship Association (Agriculture Specialist 2 2013). Established in 1979, the Carolina Farm Stewardship Association is a non-profit group that focuses on education and advocacy of local, organic foods particularly in South Carolina and North Carolina (Carolina Farm Stewardship Association [CFSA] 2014). The association is one of the oldest sustainable agriculture organizations in the Southeast. The connection between organic legislation in Georgia to the Carolina Farm Stewardship Association was cited because of the potential cross-pollination of ideas between Georgia and North Carolina. Namely, the Carolina Farm Stewardship Association may have helped foster movement by Georgia's organic farmers to push for regulation. Unfortunately, this connection could not be confirmed through interviews or through a comparison of adopted legislation, in which the legislative language varied drastically.

Vertical Diffusion

Policy conditions at the federal level played a role in the initial adoption in 2000 (SB 477) and the 2002 bill (SB 361). For initial adoption, the effect of federal policy adoption and implementation were significant but not direct factors in adoption. Specifically, in 1999 and 2000, no federal adoption occurred or federal implementation began. The effects of the vertical pressures of diffusion may stem from other activities, such as proposed rule making, but this analysis does not specifically account for the factors statistically. Nonetheless, two former GDA staff members indicated Georgia lawmakers were aware of impending federal implementation of the law but wanted to prepare and ensure a truth in labeling measure (Former GDA Staff Member 1 2013; Former GDA Staff Member 2 2013). SB 477 was intended to set up a policy framework that would allow for the state to comply with federal supremacy but also ensure, should the NOP not publish a final rule that year, that Georgia had standards in place. By 2000, skepticism had developed over the federal government's capability to actually implement a final rule. The delay of board appointments and rulemaking followed by the controversial 1997 proposed rule led to questions on whether the federal organic program would ever be implemented. As one sustainable food advocate stated, "What you sort of saw in some of that state legislation stuff was a little bit of an effort to insure that if the federal program failed there was still a process at the state level" (Sustainable Food Advocate 2013). Perhaps similar to many other states adopting organic regulation legislation from 1997–2002, Georgia's effort was an attempt to provide state level oversight of the organic market. For the 2002 bill, SB 361, the effect of federal supremacy in the policy area is a more direct effect. Indeed, the 2002 bill was adopted in response to federal implementation of the NOP that year. The amendments brought the Georgia organic law in compliance with federal law by adjusting language to be consistent with federal standards (Former GDA Staff Member 2 2013).

Issue Salience

Issue salience appears to have little effect on adoption outcomes in Georgia. Salience may have contributed to increasing consumer and producer interest, but participants cited other variables more likely to have increased overall interest and salience. In particular, the regional hubs of activity surrounding Atlanta, Savannah, and Athens are cited as the key drivers in the increasing consumer interest and demand for organic goods. These areas are characteristic (e.g., wealthier, more progressive) of what would more likely lead to the adoption of organic policy and the growth of the organic market. Increasing issue salience occurs within the time frame of initial adoption in 2000 and amending adoption in 2002, but it cannot clearly be linked to decision making at the time. Instead, the bills appear to be appeasement for and driven by a selective group within the state. Conventional farmers did not indicate any opposition to the bill and largely appear to be disinterested in the effort.

CONCLUDING GEORGIA

The development of organic policy in Georgia is a short but insightful story. As a late adopter, federal influence was one external condition motivating adoptions in 2000 and 2002. Likewise, rising consumer and producer interest in the organic market contributed to the decision for initial adoption in 2000.

Rising consumer interest in Atlanta, Savannah, and Athens gives insight for how wealth, political culture, and partisan politics played a role in adoption of state organic regulations. Furthermore, the correlation between adoption and an increase in agricultural productivity may be the result of increased attention to promoting local food within the state. The effect of salience and regionalism is less clear in this particular case. Nationally, organic issue salience was at an all time high in 2000, but no interview participant, including those associated with the development of the bill, suggested that increased media attention is what led to increased consumer interest. Similarly, several individuals referenced a regional network but no confirmation could be made if the Carolina Farm Stewards influenced Georgia's organic farmers. Evidence largely points to other factors as more significant to state legislative action.

The case of Georgia can also shed some light on why states in the Southeast were slow to adopt organic policies compared to other regions. Skepticism over organic foods may have led to delayed adoptions by any Southeastern state until 1990. Furthermore, the case also demonstrates that discussions about policy content may have been a moot point for late adopters as the co-regulatory scheme of third-party certification and registering with the state was already established by other states and at the federal level. As a late adopter, Georgia signifies that many of the variables that may have been significant to leading to adopters or middle-of-the-pack adopters are not as influential in late adopting cases. Laggards of organic legislation may be responding to gaps in federal policy as a result of consumer and industry interest in protecting the sanctity of the organic market.

NOTES

1. The "Solid South" refers to regional voting bloc behavior of states located in the Southeastern United States. Prior to realignment in the 1960s, Southern states voted predominantly for Democrats even when considering Republican and Populist movements in the region. In the mid-1960s, several Republican victories marked the end of the Democratic voting bloc behavior in the South (Tindall 1972). The cause for the "Solid South" is tied to a feeling of "deep alienation" from other parts of the country after the Civil War (Tindall 1972, 26).
2. Franciscan monks pre-date the arrival of the European settlers by approximately 160 years. The monks are responsible for the introduction of peaches in Georgia. Cherokee Indians would eventually cultivate the crop (Taylor 2013).
3. Immigration of Africans was prohibited in the colony until the slavery prohibition was repealed in 1750. Despite a ban on immigration of Africans to the colony, free and enslaved Africans still immigrated to the colony. English settlers, in particular, claimed the climate of Georgia was inhospitable for the white settlers but not for Africans.

4. Bonner (1964, 46) describes the Cotton Belt as located "in the middle portion of Georgia with its counterpart in southwestern Georgia and in the lush valleys of the former Cherokee country." After exhaustion of lands in the cotton belt, farmers moved to the southwest corner (Bonner 1964, 46 and 61).

5. The boll weevil is a pest that infests cotton crops through the deposit of eggs in fruiting structures on the plant. Yield losses occur when the eggs hatch and the boll weevil larvae feed. The boll weevil played an integral part in the history of cotton production in Georgia and the South. In Georgia, boll weevil infestations led to significant yield losses in 1914. While insecticides would provide temporary relief from infestations, Georgia began to participate in a boll weevil eradication program in 1987. The program has led to increased yields and less use of pesticides (Roberts 2013).

6. It should be noted that soil-quality problems were not just a problem in Georgia or in the Southeast. Poor soil management practices were widespread in the 1930s (Brown 2002a, 63).

7. For a current list of certified organic operations see http://apps.ams.usda.gov/nop/.

8. Many former employees of both organizations have since retired. These participants were either limited in their participation or could not be located to confirm the details.

9. The Georgia Crop Improvement Association also performs seed certification, quality assurance, food safety inspections, turf grass certification, and field inspection services. For more information see http://www.certifiedseed.org/.

10. Survey respondents were 77 percent white with 10 percent being younger than twenty-five years of age. Approximately 35 percent of respondents had annual household incomes of $35,000, which is almost $19,000 more than the average per capita income in the state in 1989.

11. The Georgia Department of Agriculture was led by Commissioner Tommy Irvin from 1969–2010 (Jones 2010). Irvin was described as being "a big conventional chicken" suggesting he extended support for conventional agricultural preferences (Former Organic Food Advocate 2013).

12. In 2001, HB 107 made non-substantive stylistic changes to Code Section 2-21-3 by changing, "Upon July 1, 2000," to "On or after July 1, 2000."

Chapter Seven

Creating Organic

This project sought to uncover why some states adopt organic food and agriculture legislation while others do not. As secondary considerations, other questions include determining how the historical development has influenced current organic practices and asking if, as a result of historical development, organic food and agriculture remains competitive in an ever-changing food market. The findings of this study reveal that a complex web of influential factors can explain state adoption of organic legislation. As such, the consequences for contemporary organic practices and consumption patterns are inadvertently affected by when and which states adopted organic policies. Leading innovative adopters, as exemplified through the California case, demonstrate the necessity for organic grassroots groups and a receptive set of political conditions to support the passage of organic regulations. However, later adopters, such as Vermont and Georgia, demonstrate variance in political conditions, agricultural production approaches, and regulatory design for the organic market. In essence, the research presented demonstrates that the adoption of organic food and agriculture regulation is dependent on the conditions of a specific time and context, which illuminate the conditions unique to the diffusion process.

A SYNTHESIS OF ORGANIC DIFFUSION

From 1976 to 2010, thirty-eight states adopted organic food and agriculture legislation. As discussed in chapter 3, there are limited explanatory factors for initial adoptions but a greater number of determinants for amending adoptions. Political culture, issue salience, and regional pressures were found as statistically significant for initial state adoption of organic regulatory legislation. Reinvention of policy had a much broader set of determinants. Federal pressures and amenable political and economic conditions were statistically significant drivers of policy change. The case studies, California, Vermont, and Georgia, support and build upon the findings from the statistical models in chapter 3. The cases provide much needed depth and meaning for how key determinants drove adoption decisions for organic food and agriculture bills over time and across space. The role of grassroots organic certification and advocacy groups were significant factors to initiating early state adopters and were highlighted in California and Vermont cases.

Internal Determinants of Diffusion

Internal determinants play a significant role in the diffusion of organic food and agriculture legislation. Eight socio-economic and political indicators were assessed to predict how state characteristics mattered for enactment decisions. Of the assumptions outlined in chapter 2, several proved to be true, or partially true, depending on the type and timing of adoption. Statistically, political culture was significant to explaining pre-1990 initial state adoptions and all amending adoptions. Furthermore, amending adoptions were also influenced by state legislative professionalism and state agricultural market activity including overall industry health and specialty crop sales. Underlying these internal conditions is the significant role of organic certification and advocacy groups. CCOF and NOFA were critical players in the development of California and Vermont legislation. In Georgia, the absence of a grassroots organic group may have delayed action amidst a backdrop of cultural skepticism toward organic practices. Collectively, the set of significant internal conditions suggest that state policy environments varied in their ability to be receptive and foster an innovative policy idea like organic food and agriculture regulations.

Grassroots Rising—The Role Organic Advocacy and Certification Groups

Grassroots organic industry groups are a significant determinant to explaining the diffusion of organic food and agriculture legislation in the United States. In California, CCOF was instrumental in the drafting of the Organic Foods Acts of 1979 and the California Organic Foods Act of 1990. For the 1990 bill, CCOF even worked with other Far West regional certifiers through WACO and collaborated with NOFA to create more standardized organic practices across the country. In addition, Congress can credit CCOF with inspiring the adoption of the OFPA. In this particular case, the role and impact of third-party certifiers and advocacy groups are direct and stems from within the state, as well as, being impacted by regional and cross-national interaction among organic third-party certifiers and advocacy groups.

A similar story is told in Vermont. NOFA-VT was the primary organizer and driver for drafting and adopting S.277. The Vermont Organic Advisory Board was a direct response to the federal efforts, which was not favorably perceived. Vermont lawmakers and organic producers were a small minority vehemently opposed to the OFPA implementation design. They supported the idea of a national law but did not support a consumer-focused law. Producers, they believed, were able to better construct the meaning of organic and could prevent the watering-down of the term. Organic certification groups were addressing fraud in the market and working to address the small differences among standards. Congressional action interrupted the efforts of the groups to reconcile any differences.

Georgia is the only case not to possess a strong presence of an organic certification or advocacy group. The Georgia Crop Improvement Association, a state entity meant to conduct a number of certifications, did play a role in promoting organic agriculture and is one of a few primary certifiers in the state. However, there is not enough evidence to suggest that Georgia Crop Improvement Association, the state's official certification entity, or any other third-party certifier operating within the state had any significant role in developing the initial legislation in 2000 or amending legislation in 2002 and 2010. Not even the influence of the Carolina Farm Stewardship, an organization that once pursued becoming a certifier, could be confirmed for a regional basis. It is likely that the absence of a grassroots organization contributed to delayed action.

Internally, organic advocacy groups were a key determinant to individual state adopters and are representative of other factors and conditions that lead to adoption that are not accounted for directly in the data analysis. Specifically, annual organic production or sales figures were not available for all thirty-five years of this study. Specialty crop sales, included as vegetable and fruit sales, can help to assess potential organic production trends. Yet, even with these proxy measurements, there is no indication for how to measure the production or demand for organic versus conventional specialty crops. The presence of organic certifiers and organizations within a state may indicate that more farmers are interested in organic production, and it is likely that consumer demand (and subsequent higher prices) may be driving that interest. Without hard data for all thirty-five years, there is no possible way to statistically test this assumption. However, the presence and analysis of organic groups in two of the cases suggests that there is some linkage between state adoption decisions and organic certifiers. The presence of organic advocacy and certification groups demonstrate why certain states and regions were likely to be leading adopters. California and Vermont both adopted before or during the critical mass and are in regions that led the nation in this innovative policy domain. The key difference in timing can be attributed to perceptions about the appropriate role of government in regulating the market, which delayed action in Vermont. In Georgia, the absence of a third-party certifier or organic advocacy group likely factored into the adoption delay and reliance on a state agency to pursue action. This scenario may be representative of what occurred in other laggard or non-adopting states.

Economic Predictors of Diffusion—Beyond Organic Groups

Several economic indicators were included in the analysis to predict if and how wealth and industry-specific factors determined state adoption decisions and diffusion. None of the determinants were found as statistically significant for initial adoptions, but the state agriculture GDP and vegetable sales were significant determinants for amendment legislation. Both determinants suggest market-driven dynamics underlying the reinvention of state policies. The effect of agriculture GDP may be interpreted as having an overall strong agricultural industry with the resources to invest into new agricultural practices and technologies. Yet, this would cautiously be interpreted until after the widespread and mainstream acceptance of organic agriculture starting in the late 1980s. Before the organic market received mainstream support, general hostility and suspicion of the fringe practice likely contributed to excess resources being directed for conventional agricultural research only.

The measures of fruit and vegetable sales, both specialty crops, were included as a proxy measure for assessing the presence and growth in organic food sales. Most early organic production centered on fruit and vegetable production. The members of CCOF and NOFA were predominantly specialty crop producers. In all three case studies, fruit and vegetable crop production was referenced as significant to their overall agriculture industry and as significant components to organic production. The significance of an industry presence is qualitatively confirmed through the presence of the organic advocacy and certification groups and statistically elaborated upon through the significance of vegetable sales to reinvention of state policies. This suggests that states with larger vegetable production sectors are more likely to pursue organic policies.

The insignificance of fruit sales should not be misinterpreted as there may be underlying production dynamics that create challenges in statistical assessment nationally. The distribution of vegetable production nationally is more equitable as vegetables can be grown across more climatic zones annually. Fruit production differs. More often fruit production is limited to particular climates and is perennial, providing less dynamic change over time.

While North Dakota or New Mexico may not generate as many fruit sales as California, Vermont, or Georgia, states like North Dakota and New Mexico can generate substantive vegetable sales given the nature of the climate and growing zone capabilities. Considering the national capacity for growing vegetables, it can be asserted that states with increasing vegetable sales most broadly signifies a greater concern for policies that would promote and protect this sector including organic food and agriculture policy. As discussed below, the production patterns of fruit and vegetable sales, as well as farm type, by state, may also shed some light on horizontal pressures for diffusion.

Ripe for the Picking—The Political Environment

The pattern of diffusion and state adoption decisions were clearly determined by political conditions within each state. All three political variables were statistically significant in some capacity. Partisan control was the weakest determinant and, to some extent, this is surprising. In California and Georgia, initial adoptions are made in years where Democrats controlled state government. In all three cases, Democrats were the sponsors of the initial bills and a majority of amending bills. However, the case studies do not illuminate a strong causal link between partisanship and adoption. Bill sponsors were likely selected based on their constituent base and the roles they served in state legislatures. In California, sponsorship of bills in 1979, 1982, and 1990 were clearly the result of constituents contacting their state assembly members for sponsorship. Both Assemblymen Michael Gage and Sam Farr were representing their constituents, predominantly members of CCOF, in the Central Valley of California when sponsoring bills. Agriculture committee members sponsored other bills. In Vermont, Elizabeth Ready's sponsorship of S.277 was grounded on a constituent-based assessment of her ability to "carry the water" (Former VOF Staff Member 2015). Likewise, Harold Ragan's sponsorship of the bill in Georgia was likely a function of his chairmanship of the Senate Agriculture Committee when the Georgia Department of Agriculture approached him about an organic law.

State legislative professionalism was significant in models examining reinvention. The comprehensive model achieved a higher threshold of significance than the OPFA-era model. In an overview of states amending legislation, more professionalized state legislatures were more likely to reinvent organic regulatory policies than less professionalized legislatures. A majority of amending adoptions occurred in states with higher than average state professionalism scores (see figure 7.1; appendix F for average state legislature professionalism scores). California, a state with one of the most professionalized state legislatures, was one of the leaders in reinventing organic regulatory policy. The state amended statutes fifteen times from 1990–2010,

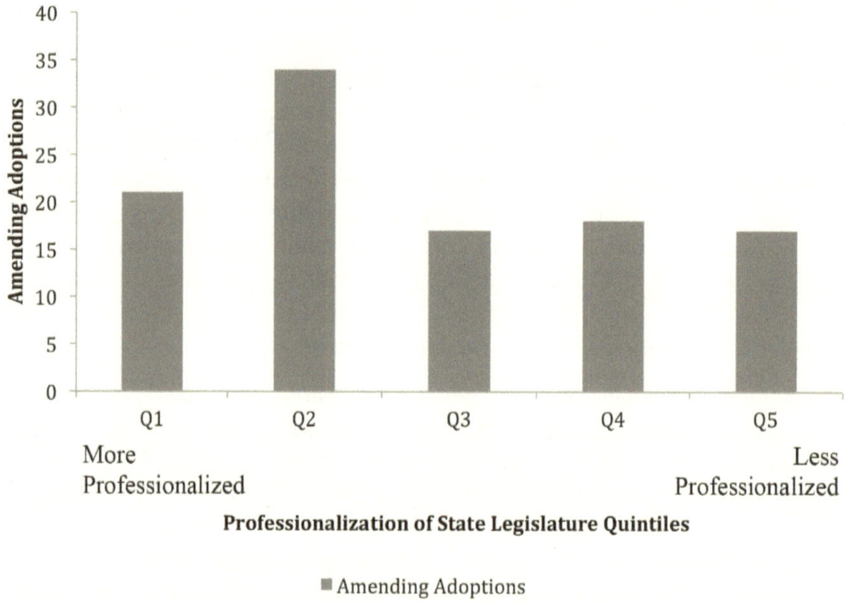

Figure 7.1. Amending Adoptions by More to Less Professionalized State Legislatures
Created by the author, based on state legislative professionalism scores obtained from Squire 2007.

by far more than any other state. California's actions are likely a significant contributor to this observed effect but collective evidence does demonstrate a higher percentage of amending adoption occurring in more professionalized state legislatures. As figure 7.1 demonstrates, there is a slight positive distribution of amending adoptions by level of state legislature professionalism. For Q1, 17 of the 21 adoptions stem from California's action; the remaining amending adoptions originate from Michigan, Pennsylvania, and Alaska. For Q2, the figure of amending adoptions is nearly double of any other quintile and stem from an even distribution of action among states with legislatures averaging professionalism ratings from .199–.248. Vermont and Georgia are not located in Q1 or Q2. Instead, both states are located in Q4. Considering the very limited number of reinvention bills adopted by either state, it is no surprise that their state legislatures are less professionalized.

The most significant political condition in this study is political culture. According to the statistical results, states with moralistic-leaning political cultures are more likely to adopt organic legislative bills, both initially, and to modify existing policy, than states with pure individualistic or traditionalistic political cultures. Moralistic political cultures demonstrate more trust in government and desire to promote policies for the public good. Adopting bills to ensure truth in labeling and protect fraud in the market supports the value of equity and fairness for both consumers and producers in society. Organic regulatory bills correct market information asymmetries.

The significance of political culture demonstrates why states like California were leading adopters while states like Georgia were laggard or even non-adopters. California's political culture is identified as moralistic with a strong individualistic strain whereas Georgia's political culture is a traditionalist culture with a strong individualistic strain. The individualistic strain in each state demonstrates government skepticism and an emphasis on individual rights. This suggests individualistic cultures may be less likely to seek government involvement for solving problems unless for the protection of individual rights. Yet, Vermont possesses a political culture that is more moralistic than California. This should indicate a higher likelihood of being an early adopter rather than a middle-of-the-pack adopter. However, the underlying currents beyond politics in the state may show a deep and trusting relationship between government and organic producers before the adoption of regulatory legislation. California's law was promoted by many organic farmers in the state for the purposes of preventing fraud in the marketplace but also to protect the market from being corrupted by conventional agriculture's interpretation of the term. While organic farmers wanted a law that would protect themselves and consumers from fraudulent claims, they were also skeptical of government and the California Department of Food and Agriculture's friendliness to conventional agricultural interests. For some organic farmers, the involvement of government would be considered heresy to their purpose and alternative production methods. However, CCOF organizers were comfortable with the idea of framing the policy as correcting an information asymmetry in the marketplace while relying on private third-party certification to ensure truth in labeling. The particular design of the 1979 and 1990 California bills demonstrate the desire to protect the public good and promote individual rights but with a deep skepticism of involving state government. The regulatory design relied on government registration of organic growers, processors, and handlers but maintained a public-private form of implementation by relying on third-party certifiers. Furthermore, the sheer length of both the 1990 and 2003 California laws also demonstrates mistrust in the rule-making process and state policy makers for developing organic industry approved standards.

The presence of both a moralistic and individualistic political nature in California may help to determine why the state is a leading innovative adopter while also explaining the co-regulatory style of policy for organic food and agriculture. The law would protect the public and organic farmers while the co-regulatory style of policy implementation would prevent government from being overly persuaded by conventional agriculture interests. If California is considered a leader in organic policy and is a significant international actor in setting environmental and consumer protection standards, other moralistic and moralistic-leaning states would take cues from California. Likewise, California may seek to emulate or mimic the behavior of other moralistic states that have strong individualistic undertones such as New York, which was the first state to adopt. Under this presumption, individualistic dominant cultures would likely be middle-of-the-pack adopters with traditionalist dominant states predominantly being laggards. The function of individualistic strains in traditionalist states, particularly in the Southeast, may serve as a gateway to considering innovative policies that are tested and successful in other states or motivated by skepticism in government capability.

Indeed, Georgia's traditionalist culture may have hindered the interest in organic food and agriculture policy until 2000 but the presence of the individualistic strain promoted policy that protects individuals. Skepticism in the federal government's adoption and the desire from the Georgia Department of Agriculture to ensure truth in labeling claims may be representative of the individualistic undertones of the moralistic culture in the state. Specifically, mistrust in the federal government in 2000 stemmed not only from the USDA's ability to develop final rules but also from a broader debate of federalism and state versus federal policy domains. States like Georgia would be skeptical of the federal government's policy development but would like to maintain a hierarchy in their own territorial boundaries by protecting the interest of consumers and industry. Thus, this individualistic strain in Southeastern states may help to explain why the region was a late adopter and demonstrates slower regional interest in organic food and agriculture policy. Similarly, the Mid-Atlantic was also a region with delayed adoption with most states being individualist dominant with strong traditionalist ties. The higher percentage of state adopters in the Mid-Atlantic region, like the Southeast, may be caused by the presence of a traditionalist culture, which needs clear proof of a policy's success or motivation to protect individual rights from unnecessary government intervention.

Vermont does not neatly fit into progression of diffusion moving from moralistic cultures first, followed by individualistic cultures, then traditionalist cultures. Vermont is closer to a pure moralistic culture than California. One possible explanation for the delayed involvement for government regulatory oversight may be tied to the very dynamics of the culture itself. NOFA-VT established a healthy working relationship with the Vermont Agency of Agriculture, Food, and Markets prior to seeking legislative support for regulations. Organic producers trusted government and the role it could serve. Seeking protection through the legislature was delayed until absolutely necessary because of a healthy working relationship already established with the bureaucracy.

This did not occur in California. CCOF was very skeptical of bureaucratic involvement and avoided the California Department of Food and Agriculture as a potential ally. Instead, the legislature became the necessary route for seeking aid and protection. The difference between the two cases illuminates the power of trust in government for timing of adoption. Mixed-moralistic cultures may result in innovative policy adoptions where trust in bureaucratic agencies is low but faith in the legislative process is high.

A Missed Cue? The Role of Universities

There was one factor that appeared in all three cases that was not expected to serve as an important factor in organic regulatory development. The role of universities was present in all three cases as either a promoter or inhibitor of organic agriculture. In California, the University of California-Santa Cruz played an instrumental role in supporting organic agriculture research and development. Today, the Center for Agroecology and Sustainable Food System at the University of California-Santa Cruz is a world-renowned educational center focusing on sustainable agriculture. Conversely, the University of Georgia is cited as receiving more support from conventional agricultural interest than support from organic research and development. In Vermont, the role of extension services in the University of Vermont is cited as unreceptive until the 1990s. Prior to this period, extension staff predominantly focused on conventional practices and dismissed the needs of organic farmers. The federal LISA program and state support for sustainable agriculture likely aided in a transition of extension service offerings.

Based on the evidence, could it be that university-based research and development is key to fostering support for organic or sustainable food systems? Moreover, could university research and development support signify an additional factor significant to organic regulatory diffusion? The role of universities in supporting particular forms of agriculture could point to delayed policy action in some states and emergence of regulations sooner in other states. The packaging of policies to support an industry and the regulations of the market may be key in the diffusion process. For example, Vermont incorporated sustainability agriculture research and education programs as part of S.277. State lawmakers found it important to cultivate production in addition to regulating the market. However, the Georgia case study identifies how limited university engagement may stifle support for an industry and appropriate regulations. While university research and education determinants were not included within the analysis, the role of universities was uncovered through the conducting of the case studies.

External Determinants of Diffusion

Internal conditions outline behavioral patterns for why individual states would be interested in pursuing organic regulatory bills. External conditions demonstrate the horizontal and vertical pressures that likely caused the spread among receptive state hosts. Statistical results proved the significance of federal action and regional pressures on diffusion in some capacity. The case studies elaborate upon these results to show how regional and national network pressures led to adoption in a number of key states first followed by more disenfranchised states and regions later. Federal action also served an underlying plot to each state's story. The adoption of OFPA in 1990 followed by a twelve-year delay in implementation led to mixed state reactions. Collectively, both horizontal and vertical pressures led to a unique pattern of state adoption behavior.

Horizontal Diffusion Pressures

Two patterns of state-to-state diffusion appear within this issue domain. The first is a regional-based diffusion pressure. The statistical models from chapter 3 find regional pressures play a significant but minimal impact role in state's initial adoption decisions both before and after adoption of the OFPA. The USDA's Agricultural Marketing Service's seven-region designation was used to group states together for analysis. The seven-region designation accounts for climatic and agricultural productivity differences among states. Furthermore, the regional identification aids in assessing factors that influence why the Southeast region has low certified organic production, which is one concern of the National Organic Program (McEvoy 2012). The results of this research suggest that regional effects matter. States can and will take cues from other states in their region regarding the adoption of new and innovative policies and programs, but the effect of cue taking is minimal at best in this scenario.

A second horizontal pattern to emerge is one created through a national network. Organic advocacy and certification groups were key drivers of early state policy adoptions. These groups were concentrated in the West, Northeast, and Upper Midwest initially (NOFA-VT Staff Member 2015; Former NOFA-VOF Staff Member 2015; Sustainable Food Advocate 2013). CCOF and NOFA-VT were identified as the drivers behind initial policies in each state. For NOFA-VT, the organization operated in relatively regional isolation until the communication channels opened nationally. The flood of state adoptions that would occur in 1985–1990 coincides with the establishment of a national network of organic farmers and producers coming in contact with one another. Moreover, the Organic Crop Improvement Association and the Organic Trade Association, both established in 1985, operated as clearinghouses of information for farmers in North America and globally. Both organizations remain significant actors in national and international organic food and agriculture policy.

The balance of both regional and national network patterns of state adoptions could be attributed to the where and how organic production occurs. As figure 7.2 demonstrates, certified organic production is primarily located outside of the Southeast as of 2007. Of significance to note is that areas with adoptions prior to 1990 have some of the largest number of organic production acres. The Northeast and Far West regions have some of the earliest adoptions and some of the largest number of certified organic acres. Furthermore, middle-of-the-pack regions such as the Rocky Mountain and North Central regions also have early pace setters and more organic production acres than the Southeast. Given the spatial evidence, Walker's (1969) original regional diffusion model may hold true. His model claimed there are regional pace setters such as California and New York, which other states within the region will follow and emulate.

The Northeast region follows Walker's predicted pattern with New York as the first legislative adopter with Connecticut and Maine following three years later. If including the dimension of organic administrative rules, Massachusetts could also be considered a follower of New York's regional pace setting. The Far West region, to the contrary, does not necessarily follow Walker's tree model. However, it is clear that regional effects matter. California was the first state to adopt organic legislation, but it was not the first state to adopt general regulations or guidelines regarding organic food labeling. Oregon initially adopted organic administrative rules in 1973. California's Organic Foods Act of 1979 closely mirrored Oregon's administrative rules both in spirit and practice. Nonetheless, it could be argued that California is a regional pacesetter for legislative adoption as not all state administrative agencies have the same sort of broad delegation of authority to develop rules. Oregon lawmakers would eventually pass legislation in 1989, but California had already begun to pioneer the policy area of organic food and agriculture with CCOF at the helms of California's efforts. CCOF had consulted with other organic industry pioneers in the Far West region including Oregon Tilth, which also alludes to the significance of industry networks within a regional construct.

In the Southeast region, the regional effects are also significant. This could be the result of delayed adoption within the region, but the Georgia case, unfortunately, cannot help further illuminate this finding and cannot lend support to Walker's regional tree-model of diffusion. A few interview participants note how Georgia, North Carolina, and Florida are more innovative in nature and would be more likely to look at each other for agricultural production cues. Moreover, there was some limited information to suggest a similar organic industry network influence between North Carolina and Georgia. The Carolina Farm Stewardship Association was identified as having probable influence on organic food and agriculture policy in the Southeast region. While the association never achieved its goal of becoming a certifier, the Carolina Farm Stewardship Association is an active promoter of local and organic food in the Carolinas.

Key interview participants indicated that organic activity is nationally situated around several hubs. Like the metropolitan hubs driving demand in Georgia, organic food production and demand also follows a spatial pattern across the United States akin to the organic production acreage presented in figure 7.2. One sustainable food advocate describes organic regionalism as the following:

> The centers of activity in my mind are California and then the Northwest, Oregon and Washington. Then you get over into Wisconsin. It's pretty strong in the Midwest. And then New England has always been a leader. Then (it goes) down to the Southeast. And then the Deep South, you know Alabama and Mississippi, Arkansas, it's barely existing. (Sustainable Food Advocate 2013)

Although this could be interpreted as regional-based diffusion, the discussion of different farm types and agricultural capabilities is also significant to determining low organic production efforts. As research on organic farming diffusion in Europe suggests farm type matters for diffusion of the practice.

As previously discussed, the measures for fruit and vegetable sales are used in this analysis as a proxy for assessing organic production and sales from 1976–2010. Assessing the sales of these two specialty crops spatially can shed additional light on how farm type contributes to regional peculiarities and an isomorphism form of diffusion. Figure 7.2 demonstrates the spatial distribution of organic production acres in 2007. Figures 7.3 and 7.4 spatially map orchard acres and vegetable acres harvested that same year. The distribution of organic farming is spread across the United States except for the Southeast. Yet the distribution of vegetable and fruit acreage is concentrated in a few key areas of the country where farm scale is relatively smaller. Namely, land in orchard acres is concentrated in the Far West, in California and Washington, and the Southeast, in Florida and southern Georgia. There are additional patches of orchard acres in Texas, Oklahoma, Michigan, and New York. It should be no surprise, given the land in orchards in California, that we found the regional significance of fruit sales in the Far West. The orchards located in the Central Valley of California are where most fruit production for the country occurs. Only Florida appears to have a comparable scale of fruit production. Comparably, vegetable acres harvested are better balanced nationally, which further contributes to our

Figure 7.2. Organic Production Acres (2007) with State Initial Adoption Date

Map created by the author based on the map from USDA National Agricultural Statistics Service Census (https://www.agcensus.usda.gov/Publications/2007/Online_Highlights/Ag_Atlas_Maps/Farms/FarmsLongDesc.php).

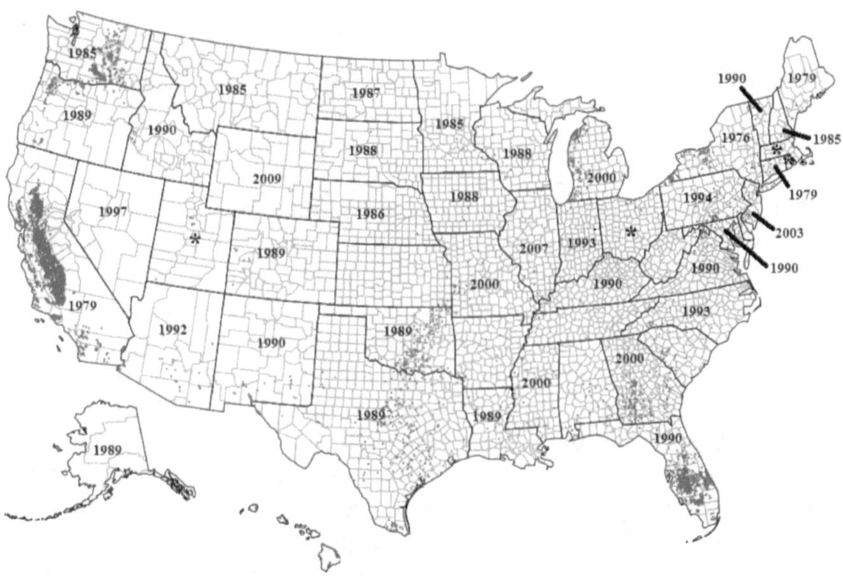

Figure 7.3. Land in Orchard Acres (2007) with State Initial Adoption Date

Map created by the author based on the map from USDA National Agricultural Statistics Service Census (https://www.agcensus.usda.gov/Publications/2007/Online_Highlights/Ag_Atlas_Maps/Farms/FarmsLongDesc.php).

Source: Base map from USDA National Agricultural Statistics Service Census (https://www.agcensus.usda.gov/Publications/2007/Online_Highlights/Ag_Atlas_Maps/Farms/FarmsLongDesc.php)

Figure 7.4. Vegetable Acres Harvested for Sale (2007) with State Initial Adoption Date
Map created by the author based on the map from USDA National Agricultural Statistics Service Census (https://www.agcensus.usda.gov/Publications/2007/Online_Highlights/Ag_Atlas_Maps/Farms/FarmsLongDesc.php).

understanding of the significance of vegetable sales to the national models of state policy diffusion. Yet the mid-section of the country and most of the Southeast remains comparatively barren from vegetable production.

Given the evidence, the question is presented on why does this spatial conception of fruit and vegetable production matter to horizontal diffusion? In part, it helps to explain how regional and national patterns of diffusion occurred. Given the differences in farm types and even scale of farming (e.g., large versus small), the diffusion of organic food and agricultural legislation may have occurred because of the concentration of fruit and vegetable production in certain states. Nationally, the ability of states like Oregon or Vermont to have large-scale farming operations growing “low-risk” crops, such as wheat and soybeans, is not the same as states in the mid-section of the country. Likewise, the ability of states in the Southeast to have the same ability to crop the vast array and amount of specialty crops as California is simply unachievable. Even Georgia has differences within the state with large-scale farming occurring in the southern half of the state and smaller-scale operations in the north. The variation in farming capabilities nationally is even evident in the patterns of agricultural lobbying. Sam Farr (2013) was quick to indicate that the politics surrounding agriculture is not so much based on partisanship but by bias in the production by state. For example, "the corn lobby is not in California or Florida but the sugar cane lobby is certainly in Florida" (Farr 2013). This would suggest that agriculture is non-partisan but grounded in particular crops and goods.

In sum, horizontal pressures of diffusion did play a significant but perhaps minimal role in state adoption decisions and diffusion. Regional pressures appear more significant to early adopters. The cross-national effect likely drove rapid diffusion in the 1980s as the organic network expanded and became more intertwined. In addition, it is likely that national networking led to states adjusting existing policies to address discrepancies in acceptable practices.

Vertical Diffusion Pressures

Federal intervention into a burgeoning organic policy domain fundamentally altered the diffusion pattern. OFPA punctuates the diffusion process at a critical moment in 1990 and then is followed by a twelve-year period of uncertainty until the NOP was officially implemented in 2002. Both federal adoption and federal implementation were significant to motivating state legislative action. For federal adoption, the passage of the national Organic Foods Production Act of 1990 coincided with a critical mass period of state legislative adoptions. According to individuals within California and Vermont, there was no guarantee that a federal law would be adopted; state efforts were operating on a parallel track of development and passage to the federal law. For California, the reception of federal preemption may have been more positive, as CCOF staff member Mark Lipson was clearly linked to federal legislation development. In Vermont, the direction of the federal law reported by Kathleen Merrigan was worrisome and led to a state policy design emphasizing producers and not consumers. Other states likely operated under similar assumptions as each sought to initiate or reinvent state policy. In essence, some state lawmakers and key constituent groups may have perceived federal efforts more positively than others.

This research also considered how delayed implementation of federal law impacted state adoption decisions. There are far fewer initial state adoptions in immediate years following 1990. States may have been waiting for the development of rules by 1993. However, after the deadline came and passed with no rules being developed and delayed appointments to the NOSB, state laws remained regulators of the market. The continuation of initial state adoptions after 1993 signifies the degree of uncertainty regarding when and if federal implementation would occur. The controversial 1997 proposed rules did not instill confidence for many. Georgia, as a lagging adopter, was inspired to action partially over concerns about the federal law not being implemented. In California, repetitive statutory adjustments in the 1990s showed a persistent concern for if and when the NOP would be implemented. Vermont's response was slightly different. Only after the first proposed rule in 1997 was released did the state act. The response did not substantively change existing statutes but was a resolution critiquing the NOSB and proposed NOP rules. The time period between federal adoption and implementation clearly marks a period of uncertainty for states in making adoption decisions.

Once the National Organic Program's final rule went into effect in 2002, the number of new innovators drastically slowed and many state adopters amended but did not terminate state statutes. The decline of enacted state statutes is minimal with most states maintaining organic food and agriculture statutes through 2010. A majority of states saw no reason to terminate existing policy but instead modified language to accommodate federal preemption. Both California and Georgia's amended legislation in 2002 altered existing state statutory language to comply and account for federal law. For California, the need to continually amend legislation after 2002 was to challenge federal law. In particular, California lawmakers took aim at the NOP for not sufficiently regulating organic aquaculture in 2005 or organic fertilizer content in 2009. Both actions followed the 2003 challenge to regulate organic cosmetics. Unlike the 1990 adoption, California law was not operating on a parallel track but was instead intentionally seeking to reshape acceptable practices and standards. California's effect on federal law is just as significant as federal law's effect on California. While the case of California may be the exception rather than the rule, the efforts of California lawmakers suggests dynamic and tense intergovernmental relations. Federal law may be the supreme law of the land, but states are willing to develop their own policies in light of gaps in federal policy. For California, the discrepancies in the federal law go beyond uncertainty of federal adoption or implementation. It extends to challenging the permissible federal practices and standards.

For a very small number of states, challenging the federal government or maintaining an organic law was not of interest. Nine states terminated existing statutory language from 2001–2010. For those states that terminated state statutes, the cause may be related to a desire to permit federal supremacy or to remove an outdated law. New York terminated its 1976 law in 2010. It was considered an "outdated law" and no longer needed (New York Civil Servant 2013). For Vermont, the 2002 statutory termination was the removal of an outdated law in the sense that federal supremacy was imminent. There is no evidence to suggest the legislature ever considered adjusting statutory language to accommodate federal law.

Vertical determinants affected the action of states in the diffusion and reinvention of organic food and agriculture regulations. Federal adoption and implementation each shaped perceptions about the role of states in regulating the organic market. States were forced to accommodate their own policies with the events occurring within Congress and rules established by the NOSB. Yet, states also sought to push back against undesirable federal implementation design and rules. California and Vermont present two cases with stories of lawmakers that disagree with the direction of the federal law and efforts to regulate the organic market. For this policy domain, states were as equally influenced by federal action as the federal government was influenced by states. States served as little laboratories of democracy and then sought to continue shaping organic regulatory law once it percolated upwards to the federal level.

Issue Salience

National attention to organic food and agriculture was minimal until the late 1980s. Then a series of focusing events for both consumers and producers suddenly spotlighted the issue. News coverage ranged from pointing out the dangers of conventional agriculture, where to buy organic produce, and stories of fraud. Each story highlighted the organic market and potential benefits and pitfalls to participation. In California, the passage of the Organic Foods Act of 1979 was in response to events surrounding the increasing use of the term natural and a series of stories about fraud in the organic market. While these stories were not in the national news, organic industry insiders were spurred to action by local reports about fraud and concerns about conventional agriculture co-opting the term natural. By the late 1980s, the California carrot caper scandal and the Alar scare are two focusing-event stories that enhanced pressure on revising the law in California and initiating a law in Vermont. Market fraud was a continual concern, but the amount of attention given to the carrot caper scandal and the Alar scare truly elevated the salience of organic food and agriculture issues. Elevated levels of attention to organic agriculture were positively associated with state lawmakers decision to adopt bills both initially and to reinvent existing law.

For the twenty years following the adoption of the Organic Foods Production Act of 1990, issue salience continued to increase until 2007, when the number of stories began to decline. Stories addressed fraud in the marketplace, concerns about conventional agriculture, and the benefits of eating organic, but there were also a number of stories specifically addressing the development of federal organic policies, emerging new organic products (e.g., aquaculture and apparel), and how organic food compares to other sustainable food practices and labeling schemes such as local, natural, fair-trade, non-GMO, and community-based agriculture. The decline of issue salience does not appear to have an adverse impact on diffusion, as the process was nearly complete, but the decline does coincide with a few states terminating existing statutes and pursuing other sustainable food and agriculture policies such as local state-based food programs.

Issue salience played a positive role in motivating states to adopt and reinvention state regulatory policies. The effect was by far the largest for early adopters rather than later adopters. The decline in salience recently may highlight other concerns such as the long-term viability of organic food and agriculture policy at the state level. States and consumers may be more interested in newly emerging market approaches such as local, natural, or GMO-free goods.

Understanding Organic Diffusion

The diffusion of organic food and agriculture legislation is determined by a complex combination of factors. State adoption decisions were truly based on the complex interaction among internal state characteristics, the actions of other states and connectivity of organic networks nationally, federal actions, and elevated attention to organic food and agriculture. Over time the conditions significant for initial adoption varied. Early adopters were driven by increasing attention to the organic issue, regional dynamics, more amenable political culture, and the presence of organic producer organizations (e.g., moralistic-leaning). For those states adopting during the critical mass, the perfect storm of federal interest in regulations and the emergence of a nationally connected network conjoined with issue salience, regional pressures, and particular political cultures to spur diffusion. After 1990, lagging state adopters were influenced by a different set of conditions. Namely, laggards were spurred to action by delayed federal action and failure of the federal government to appropriately regulate the market. Laggard adopters faced less risk in pursuing their own regulatory policies.

Reinvention of organic legislative policies was also driven by a dynamic set of determinants. Reinvention was the result of federal actions, including the delay of implementation and the NOP's final rule implementation in 2002, and a host of internal conditions permitting interest and ability to refine statutory language. Over time, the reinvention of state policies reflects emerging concerns in the market. California's major overhaul in 1990 points to rising concern of fraud and interest in the market. In the 1990s, state lawmakers made adjustments for delayed implementation and addressed other discrepancies in standards including appropriate transitionary periods. Accommodation for federal supremacy would occur in 2002 in California, Georgia, and Vermont. California would be the only state to continue, however, to respond to federal regulations. Several amending adoptions pushed the boundaries for organic food and agriculture policy to clearly address organic cosmetics, aquaculture, and organic fertilizer inputs. The federal government delayed or was lax in covering rules for these aspects of organic production.

THE FUTURE OF ORGANIC

Organic food and agriculture regulations began in U.S. states and has weathered through five decades of state and federal action. Over time the design and content of policy has shifted but stayed true to the spirit of early organic producers and maintained a co-regulatory approach. This book provides clarity to our understanding of how organic food and agriculture regulations emerged via legislation and, in turn, it demonstrates the path dependent nature of current national standards. The organic market and national regulation of the market face several challenges as consumer preferences change and producers shift to meet those demands. Some of the current challenges may or may not easily be resolved or addressed given the historical patterns of interest and adoption of organic regulatory activity.

One of the current challenges in the organic market is "geographic-specific challenges" discussed by Miles McEvoy (2012). The historical evidence documented within this book sheds some light on the regional variation in certified organic production and interest in organic food. The Southeast, the region of primary concern, did not possess the same political or economic conditions necessary for state innovation in organic regulatory policy. As a generally conservative area, organics are perceived as a "California thing." In addition, there may be a stronghold of conventional agriculture research and development interest in the region, as well. The legacy of such perceptions and production patterns may have become more challenging once federal law was enacted and embedded into current constructs of the consumer-oriented approach of regulation today.

The current federal organic policy is described as consumer-oriented in nature rather than producer-oriented. Some scholars refer to this in other terms such as product versus process based (Clark 2015) or spreaders versus tillers (Obach 2015). No matter the label, the idea of a consumer-oriented policy encompasses assumptions of the watering down of the term "organic" and rise of large agribusinesses and farms into the market. The philosophies and principles of small, local farms inherently disappear in the consumer-based approach according to critics of the current policy approach. The rules set forth by the NOP narrowly define organic practices and omit a list of key principles such as soil health, humane and ethical treatment, and environmental or ecological harmony. The conflict over the current policy design is representative of only a small sect of opinions in a number of key adopting states. It appears that only certain perceptions and values were translated into action. Namely, the voices of organic certifiers on the west coast and, to a lesser extent, the east coast were heard the loudest. State lawmakers in these regions funneled the co-regulatory implementation design and normative organic practices into legal standards, which later transferred to the national level.

The end design of the federal policy is a clear representation of the efforts made among those early innovators. In the case of California, the effect of the state policy actions still shapes federal activity today. The consumer-oriented approach is derived from the state actors who later worked at the national level. Even those within states who opposed the consumer-oriented approach still participated in the national organic policy development. Only lagging innovators, largely those in the Mid-Athlantic and Southeast, were not present in developmental conversations. Indeed, only one member of NOSB prior to 2002 was from the Southeast, and there is an overrepresentation from certain areas of the country with more organic production (NOSB, n.d.). None of this is to suggest that the current law design was the result of intentional discrimination, but it is clear that certain regional and state voices were given a seat at the federal policy table.

For some states, the lack of representation on the NOSB may not be a large concern. Changing market dynamics point to other policy concerns. For example, while some Southern states are skeptical of organic agriculture, they are also more willing to support local food and local agricultural policies. Policy makers in Georgia and Kentucky established local, state-pride food programs. In the case of the latter, the Kentucky Proud program was authorized when state organic regulations were terminated. Similar local, state-pride food programs have also resonated elsewhere including Minnesota, Indiana, Alaska, Pennsylvania, Michigan, Missouri, and Arkansas. Some states, like Minnesota in 1990, were early adopters of such programs while others are new to the idea. The premise of all of the programs includes promotion of state produced and grown goods with regulations to oversee compliance in marketing. It is a similar marketing policy design to "organic" but with a focus on products produced in a state.

The sudden interest of state lawmakers in local, state food programs coincides with one of several emerging food trends that increased in prominence as organic food salience peaked and began to decline in popularity in 2008. Local foods and community-based agriculture were the predominant markets swaying attention away from organic foods toward newer innovative approaches. In the case of Vermont and California, organic food production has always been intimately intertwined with community-supported agriculture and local production. The market today still values these marketing dynamics, and governments are seeking to create regulations for other market dynamics. Today, state and federal governments have or are considering legislative measures to regulate other alternative markets, including natural foods and the identification of GMOs. Vermont became the first state to adopt a non-contingency based GMO-labeling law in 2014. Connecticut and Maine adopted contingency-based GMO-labeling laws in 2013 and 2014 respectively. To date, seventeen other states have proposed legislation to required GMO identification and Congress passed a 2016 bill mandating GMO labeling. For the federal law, critics are quick to point out that "nobody's super happy about it" given the number of compromises (Charles 2016). The labeling of GMO goods challenges the organic market as some consumers may be willing to accept conventionally grown foods that are GMO-free.

Labeling of natural foods is also seeing a revived interest. It's been over thirty-seven years since the FTC's failed attempt and Connecticut's successful enactment of regulations on natural food labeling, but consumers and manufacturers are now demanding government involvement with regulating the term. In November 2015, the Food and Drug Administration opened a public comment period seeking input on if and how the agency should regulate the natural food label (80 FR 69905). Consumer confusion and deception is largely cited as a concern. Companies can benefit from misinterpretation of the natural label. This is what drove early organic advocates to seek regulation.

It is clear the density of the alternative food market is increasing as states and federal lawmakers continue to regulate other alternative markets. While market sales for organic products still grows, there is the potential threat of infringement of other food labeling schemes luring customers away from organic goods. Food can be identified as local, natural, GMO-free, fair trade, environmentally certified, or stemming from community-support agriculture in addition to the current regulated term of organic. Some consumers may stay within the organic market while others may be attracted to a different market for financial or personal reasons. How then can organic remain relevant in the ever-competitive food market? In addition, is there any potential for the certified organic market to expand and become relevant in the Southeast? It is hard to say with precise certainty. If the evidence from Georgia holds true, local food programs could be a gateway for consumers in the Southeast to explore other alternative food production systems. Consumer exploration into where their food is sourced and how it's grown may contribute to growing curiosity in organic, natural, fair trade, or non-GMO products. Yet criticism of the organic market goes beyond preconceived notions of it being a "California thing." On a national scale, the standards and cost associated with becoming certified organic through the NOP may not be worth the investment for some growers and producers. There may be more value in seeking local, natural, fair trade or non-GMO certification to serve different consumer markets. While it is not the intention of this analysis to diverge into the development of the organic prohibited substance list or other organic standards, there remain significant debates about if and how the original philosophies of early organic farmers translates into modern-day policy standards. Given changing market dynamics and vocal critics, the NOSB and NOP appear to be listening. In April 2016, the USDA announced plans to strengthen humane animal treatment regulations for organic meat, egg, and dairy products. The proposal was a clear indication to meet consumer expectations for organic label integrity while also addressing producer complaints of an uneven playing field (AMS 2016). The development of these rules may signify an end to a narrowly defined consumer-approach to regulation. Only time can determine if and how other core philosophies may be embedded and altered into current national standards.

As the organic market continues to evolve, it should not be forgotten that "organic" is now a government regulated term with a very precise meaning based on production standards set by the NOP. Therefore, if organic growers and producers seek to remain relevant, it may be wise to reassess the standards currently in place and to also consider dovetailing with other alternative food markets. It was the intention of this book to demonstrate how current predicaments may be dependent and explained by the actions of U.S. states governments. The historical development of organic policy in the United States reflects the values and policy preferences set by organic fruit and vegetable growers in the 1970s centered in the Far West, Upper Midwest, and Northeast. Dissenting opinions or opinions outside of these regions may have largely been ignored in the process, thereby leading to current challenges present in the organic market. If one is to know the future of organic, you may want to look at developments in the past.

Appendix A

Independent Variables

Variable	*Measurement*	*Source of Data*	*Expected Relationship*
Partisan Control	-1 = Republican Control 0 = Neither Party in Control 1 = Democratic Control	Klarner (2013)	Democratically Controlled Legislatures = More Likely to Adopt
State Legislative Professionalism	Interval (0–1) *0 = Less Professional* *1 = More Professional*	Squire (2007)	More Professionalized Legislatures = More Likely to Adopt
Political Culture	Elazar's Culture as measure............ Interval (1–9) *1 = Perfectly Moralistic* *5 = Perfectly Individualistic* *9 = Perfectly Traditionalist*	Elazar (1972); Sharkansky (1969)	Moralistic-leaning Cultures = More Likely to Adopt
Per Capita Income	Interval ($) Per Capita Income of State Residents	Bureau of Economic Analysis	Higher Levels of Income = More Likely to Adopt
State Agriculture GDP (logged)	Interval (0–100%) Percentage of state's GDP that results from the agriculture industry	Bureau of Economic Analysis	Higher GDP from Agriculture = More Likely to Adopt

(continued)

Variable	*Measurement*	*Source of Data*	*Expected Relationship*
Fruit Sales (logged)	Interval (0–100%) Percentage of state agricultural market value sales resulting from fruit, tree nuts, and berries	USDA Census of Agriculture (1964–2007)	Higher Fruit Sales = More Likely to Adopt
Vegetables Sales (logged)	Interval (0–100%) Percentage of state agricultural market value sales resulting from vegetables, melons, potatoes and sweet potatoes	USDA Census of Agriculture (1964–2007)	Higher Vegetable Sales = More Likely to Adopt
Presence of Organic Certification/ Farmers Organizations	Qualitatively Assessed	Archival Research; Databases	Presence Increases Likelihood of Adoption
Issue Salience (frequency)	Interval # of articles published in *The New York Times* in a year	*New York Times* Print Edition	Higher Salience = More Likely to Adopt
Federal Adoption	Binary Federal adoption of organic legislation (initial or amending)	Archival Research	Federal Adoption = More Likely to Deter States Initial/ Spur Amendments
Federal Implementation	Binary Federal statutes officially enacted/ in effect	Archival Research	Implementation = More Likely to Amend
Regional Adoption	Percent of states within region having adopted (USDA AMS 7-Region Designation)	USDA AMS Regional Assessment	Higher % = More Likely to Adopt

Source: Compiled by author.

Appendix B

Univariate Results

Variable	*Obs*	*Mean*	*Std. Dev.*	*Min*	*Max*
Any Adoption	1750	.081	.274	0	1
Initial Adoption	1750	.021	.145	0	1
Amending Adoption	1750	.06	.237	0	1
Partisan Control	1750	.152	.652	-1	1
State Legislative Professionalism	1750	.197	.120	.027	.659
Political Culture	1750	5.06	2.56	1	9
Per Capita Income	1750	19,273	4,044	10,091	35,218
Log Agriculture GDP	1750	7.03	1.06	3.72	9.77
Log Ag. Sales from Fruit	1750	9.50	2.54	.693	16.3
Log Ag. Sales from Veggies	1750	10.2	2.01	4.18	16.1
Issue Salience (Logged)	1750	2.57	1.53	0	5.01
Federal Adoption	1750	.228	.420	0	1
Federal Implementation	1750	.257	.437	0	1
Regional Adopt Percentage	1750	.418	.295	0	.857

Source: Data compiled by the author; See appendix A.

Appendix C

California Presidential Election Results by County (2000–2012)

Figure C.1. California Presidential Election Results by County, 2000. Gray counties voted Democratic; white counties voted Republican.

Created by the author with data from the California Secretary of State (n.d.).

Figure C.2. California Presidential Election Results by County, 2004. Gray counties voted Democratic; white counties voted Republican.

Created by the author with data from the California Secretary of State (n.d.).

Figure C.3. California Presidential Election Results by County, 2008. Gray counties voted Democratic; white counties voted Republican.

Created by the author with data from the California Secretary of State (n.d.).

Figure C.4. California Presidential Election Results by County, 2012. Gray counties voted Democratic; white counties voted Republican.

Created by the author with data from the California Secretary of State (n.d.).

CALIFORNIA WEALTH AND POPULATION DENSITY

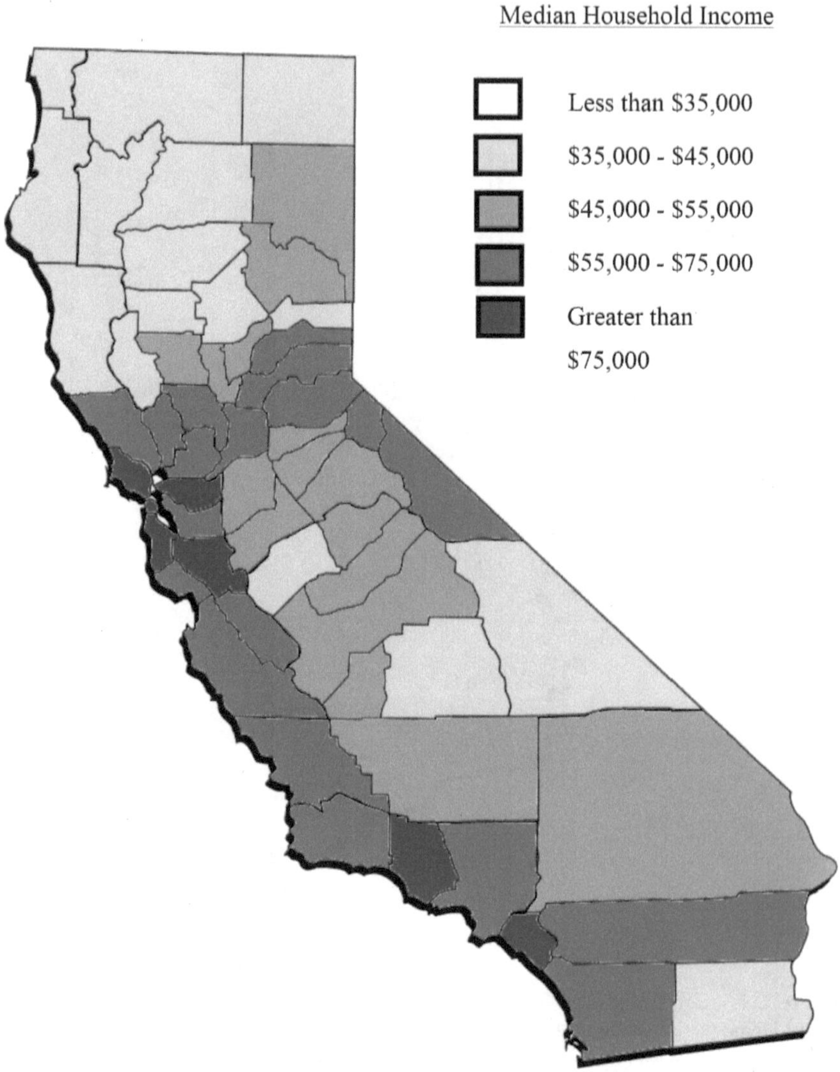

Figure C.5. California Wealth Distribution
Created by the author with information obtained from the U.S. Census 2010.

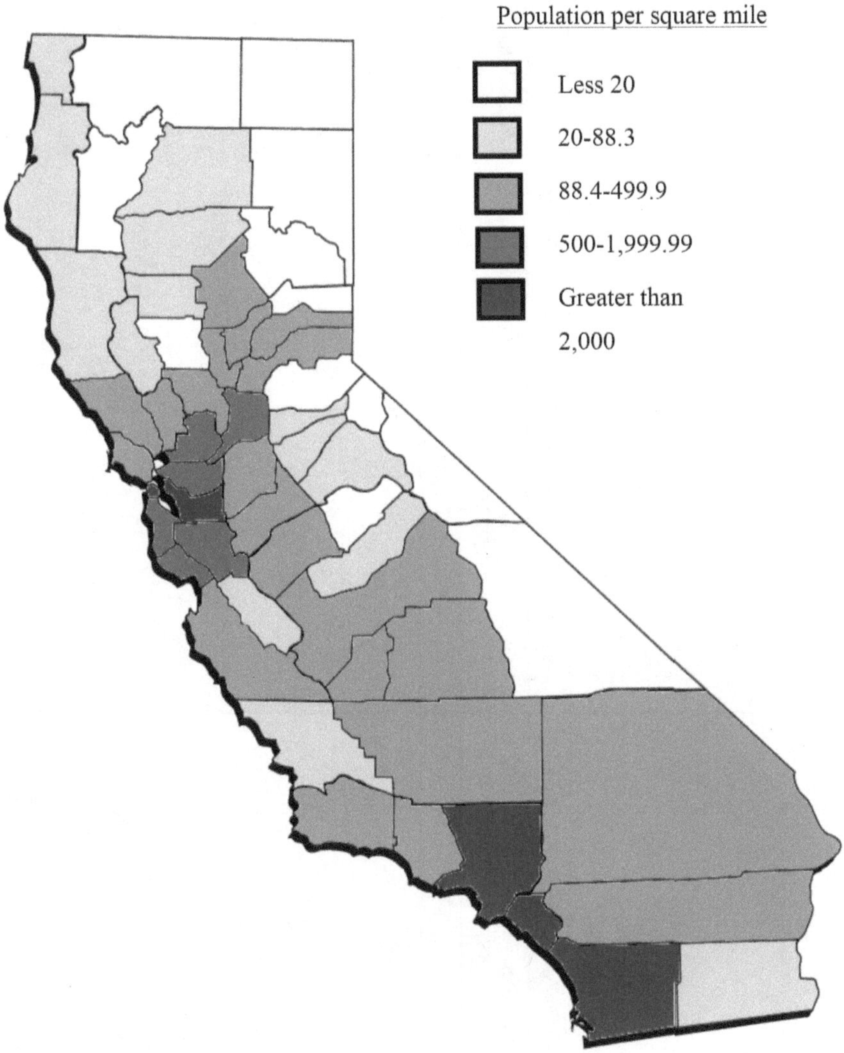

Figure C.6. California Population Distribution
Created by the author with information obtained from the U.S. Census 2010.

Table C.1. West Region Political Demographics

	State Legislature Professionalism Ranking (PR), State Government Partisan Control (GPC), and Political Culture (PC)											
	Alaska (PC 6.0)		*California (PC 3.55)*		*Hawaii (PC 8.25)*		*Nevada (PC 3.66)*		*Oregon (PC 2.0)*		*Washington (1.66)*	
Year	*PR*	*GPC*	*PR*	*GPC*	*PR*	*GPC*	*PR*	*GPC*	*PR*	*GPC*	*PR*	*GPC*
1979	.320	S	.526	D	.246	D	.130	S	.233	S	.212	S
1980	.319	S	.540	D	.250	D	.132	S	.226	S	.215	S
1981	.317	S	.554	D	.255	D	.135	S	.219	S	.217	R
1982	.316	S	.568	D	.259	D	.137	S	.212	S	.220	R
1983	.315	S	.583	S	.263	D	.139	D	.204	S	.222	S
1984	.314	S	.597	S	.267	D	.141	D	.197	S	.225	S
1985	.312	S	.611	S	.272	D	.144	S	.190	S	.227	D
1986	.311	S	.625	S	.276	D	.146	S	.183	S	.230	D
1987	.303	S	.620	S	.274	D	.148	S	.180	D	.227	D
1988	.295	S	.614	S	.271	D	.149	S	.177	D	.224	D
1989	.287	S	.609	S	.269	D	.151	S	.174	D	.220	S
1990	.279	S	.603	S	.266	D	.152	S	.171	D	.217	S
1991	.272	S	.598	S	.264	D	.154	S	.168	S	.214	S
1992	.264	S	.592	S	.262	D	.155	S	.165	S	.211	S
1993	.256	S	.587	S	.259	D	.157	S	.162	S	.208	D

1994	.248	S	.581	S	.257	D	.158	S	.159	S	.204	D
1995	.240	S	.575	S	.254	D	.160	S	.156	S	.201	S
1996	.232	S	.570	S	.252	D	.161	S	.153	S	.198	S
1997	.231	S	.578	S	.248	D	.158	S	.154	S	.198	S
1998	.231	S	.586	S	.244	D	.154	S	.155	S	.198	S
1999	.230	S	.594	D	.240	D	.151	S	.156	S	.198	S
2000	.229	S	.602	D	.237	D	.148	S	.156	S	.197	S
2001	.228	S	.610	D	.233	D	.145	S	.157	S	.197	S
2002	.228	S	.618	D	.229	D	.141	S	.158	S	.197	D
2003	.227	R	.626	D	.225	S	.138	S	.159	S	.197	S
2004	.225	R	.619	S	.231	S	.142	S	.161	S	.200	S
2005	.224	R	.611	S	.237	S	.145	S	.163	S	.202	D
2006	.222	R	.604	S	.244	S	.149	S	.166	S	.205	D
2007	.220	R	.596	S	.250	S	.152	S	.168	D	.207	D
2008	.219	R	.589	S	.256	S	.156	S	.170	D	.210	D
2009	.217	S	.581	S	.262	S	.159	S	.172	D	.212	D
2010	.215	S	.574	S	.268	S	.163	S	.174	D	.215	D

PR (State Legislature Professionalism) 0 = least professionalized; 1 = most professionalized
GPC (Partisan Control of State Government) D = Democrat; S = Split/Divided; R = Republican

Source: Data compiled by author from various sources; see appendix A.

Table C.2. AB 2823 Floor Votes

*Floor Votes on AB 2823 by Member and Party**				
Floor Vote/Date	*Ayes*		*Nays*	
Senate Floor/ August 20, 2002	Alarcon [D]	O'Connell [D]	Ackerman [R]	Poochigian [R]
	Alpert [D]	Ortiz [D]	Battin [R]	
	Bowen [D]	Peace [D]	Dunn [D]	
	Burton [D]	Perata [D]	Haynes [R]	
	Chesbro [D]	Polanco [D]	Johannessen [R]	
	Costa [D]	Romero [D]	Johnson [R]	
	Escutia [D]	Scott [D]	Knight [R]	
	Figueroa [D]	Sher [D]	Margett [R]	
	Karnette [D]	Soto [D]	McClintock [R]	
	Kuehl [D]	Speier [D]	Monteith [R]	
	Machado [D]	Torlakson [D]	Morrow [R]	
	McPherson [R]	Vasconcellos [D]	Oller [R]	
	Murray [D]			
Assembly Floor/ August 28, 2002	Alquist [D]	Matthews [D]	Aanestad [R]	Pescetti [R]
	Aroner [D]	Migden [D]	Ashburn [R]	Richman [R]
	Briggs [R]	Nakano [D]	Bates [R]	Runner [R]
	Calderon [D]	Nation [D]	Bogh [R]	Strickland [R]
	Canciamillia [D]	Negrete-McLeod [D]	B. Campbell [R]	Wyland [R]
	Cardenas [D]		J. Campbell [R]	Wyman [R]
	Cardoza [D]	Oropeza [D]	Cogdill [R]	Zettel [R]
	Chan [D]	Pavley [D]	Cox [R]	
	Chavez [D]	Reyes [D]	Daucher [R]	
	Chu [D]	Salinas [D]	Dickerson [R]	
	Cohn [D]	Shelley [D]	Harman [R]	
	Corbett [D]	Simitian [D]	Hollingsworth [R]	
	Correa [D]	Steinberg [D]	Kelley [R]	
	Dutra [D]	Strom-Martin [D]	La Suer R]	
	Firebaugh [D]	Thomson [D]	Leach [R]	
	Florez [D]	Vargas [D]	Leonard [R]	
	Goldberg [D]	Washington [D]	Leslie [R]	
	Liu [D]	Wayne [D]	Maldonado [R]	
	Longville [D]	Wesson [D]	Mountjoy [R]	
	Lowenthal [D]	Wiggins [D]	R. Pacheco [R]	
	Maddox [R]	Wright [D]	R. Pacheco [R]	

**Political Party – [D] = Democrat; [R] = Republican*

Source: Data collected from the California Legislation Information website.

Appendix D

Vermont Wealth and Population Density

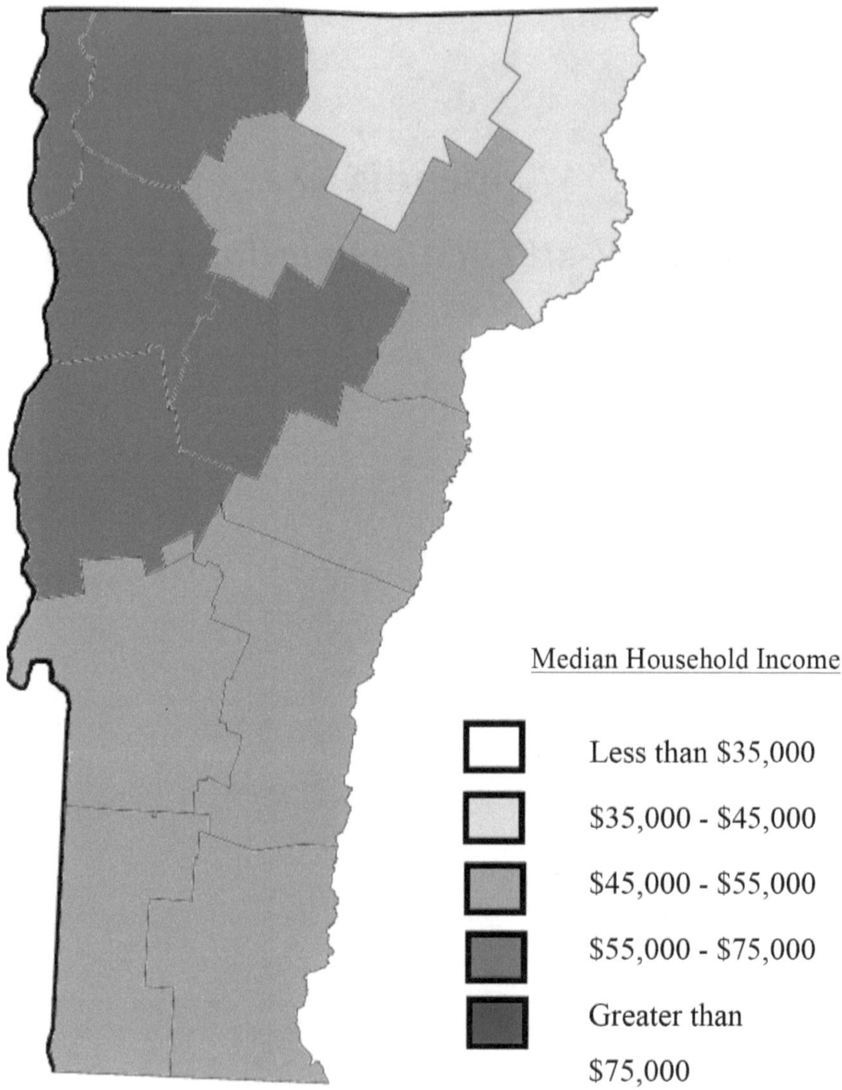

Figure D.1. Vermont Wealth Density
Created by the author with data from the U.S. Census 2010.

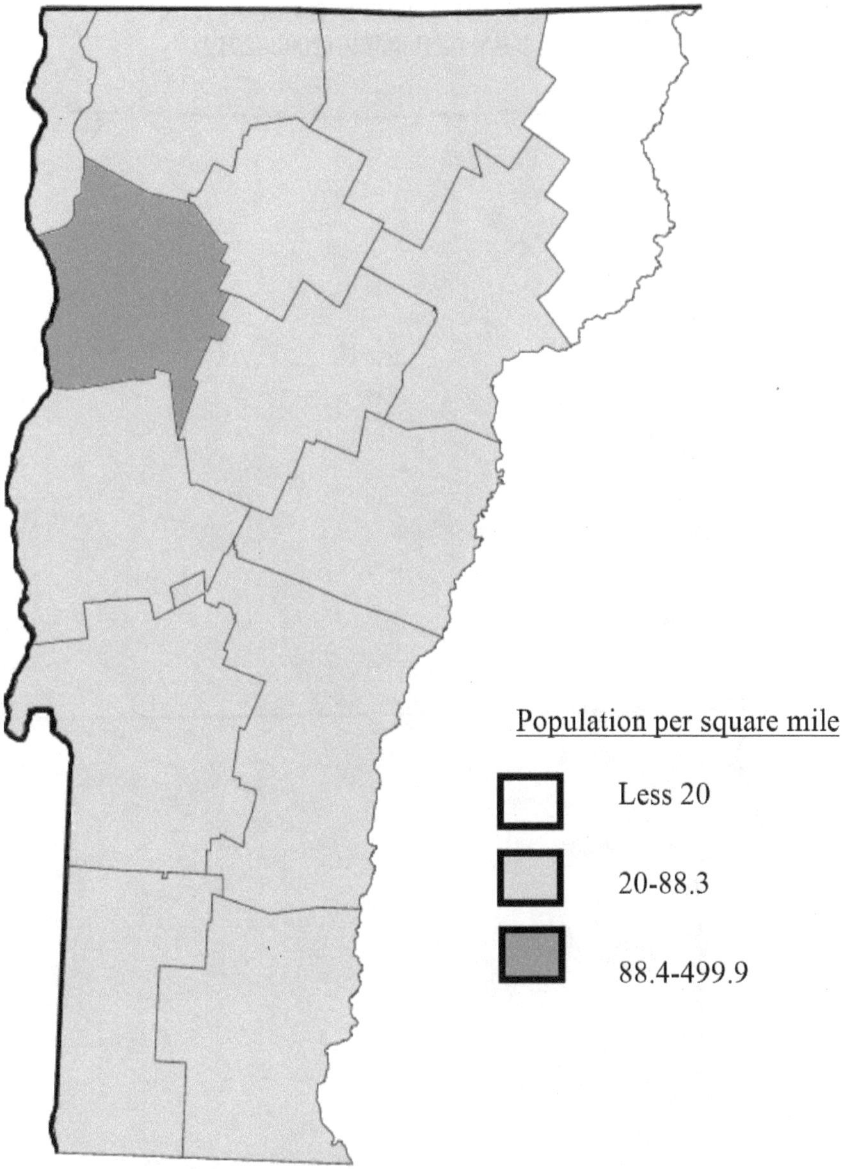

Figure D.2. Vermont Population Density
Created by the author with information obtained from the U.S. Census 2010.

VERMONT PRESIDENTIAL ELECTION RESULTS BY COUNTY (2000–2012)

Figure D.3. Vermont Presidential Election Results 2000. Gray counties voted Democratic; white counties voted Republican.

Created by the author with data from the Vermont Secretary of State (n.d.).

Figure D.4. Vermont Presidential Election Results 2004. Gray counties voted Democratic; white counties voted Republican.

Created by the author with data from the Vermont Secretary of State (n.d.).

Figure D.5. Vermont Presidential Election Results 2008. Gray counties voted Democratic; white counties voted Republican.

Created by the author with data from the Vermont Secretary of State (n.d.).

Figure D.6. Vermont Presidential Election Results 2012. Gray counties voted Democratic; white counties voted Republican.

Created by the author with data from the Vermont Secretary of State (n.d.).

Table D.1. Northeast Region Political Demographics

	State Legislature Professionalism Ranking (PR), State Government Partisan Control (GPC), and Political Culture (PC)													
	Connecticut (PC 3.0)		*Maine (PC 2.33)*		*Massachusetts (PC 3.66)*		*New Hampshire (PC 2.33)*		*New York (PC 3.62)*		*Rhode Island (PC 3.0)*		*Vermont (PC 2.33)*	
Year	*PR*	*GPC*	*PR*	*GPC*	*PR*	*GPC*	*PR*	*GPC*	*PR*	*GPC*	*PR*	*GPC*	*PR*	*GPC*
1976	.186	D	.170	S	.404	D	.072	S	.387	S	.148	D	.134	S
1977	.190	D	.173	S	.398	D	.069	S	.394	S	.146	D	.133	S
1978	.195	D	.177	S	.392	D	.065	S	.400	S	.144	D	.131	S
1979	.200	D	.180	S	.386	D	.062	S	.407	S	.142	D	.130	R
1980	.205	D	.175	S	.419	D	.059	S	.443	S	.143	D	.132	R
1981	.209	D	.171	S	.451	D	.056	S	.479	S	.144	D	.134	R
1982	.214	D	.166	S	.484	D	.053	S	.515	S	.145	D	.136	R
1983	.219	D	.161	D	.516	D	.051	R	.551	S	.145	D	.139	R
1984	.224	D	.156	D	.549	D	.048	R	.587	S	.146	D	.141	R
1985	.228	S	.152	D	.581	D	.045	R	.623	S	.147	S	.143	S
1986	.233	S	.147	D	.614	D	.042	R	.659	S	.148	S	.145	S
1987	.228	D	.142	S	.586	D	.041	R	.645	S	.145	S	.142	S
1988	.222	D	.137	S	.558	D	.040	R	.630	S	.141	S	.139	S
1989	.217	D	.132	S	.529	D	.040	R	.616	S	.138	S	.137	S
1990	.211	D	.127	S	.501	D	.039	R	.601	S	.134	S	.134	S
1991	.206	S	.123	S	.473	S	.038	R	.587	S	.131	D	.131	S

1992	.200	S	.118	S	.445	S	.037	R	.573	S	.127	D	.128	S
1993	.195	S	.113	S	.417	S	.036	R	.558	S	.124	D	.125	S
1994	.189	S	.108	S	.388	S	.036	R	.544	S	.120	D	.123	S
1995	.184	S	.103	S	.360	S	.035	R	.529	S	.117	S	.120	S
1996	.178	S	.098	S	.332	S	.034	R	.515	S	.113	S	.117	S
1997	.180	S	.097	S	.340	S	.033	S	.510	S	.116	S	.121	D
1998	.181	S	.095	S	.347	S	.032	S	.505	S	.119	S	.125	D
1999	.183	S	.094	S	.355	S	.031	S	.500	S	.122	S	.129	D
2000	.185	S	.093	S	.362	S	.030	S	.495	S	.124	S	.132	D
2001	.187	S	.092	S	.370	S	.029	S	.490	S	.127	S	.136	S
2002	.188	S	.090	S	.377	S	.028	S	.485	S	.130	S	.140	S
2003	.190	S	.089	D	.385	S	.027	R	.480	S	.133	S	.144	S
2004	.191	S	.089	D	.368	S	.028	R	.501	S	.133	S	.138	S
2005	.192	S	.089	D	.350	S	.028	S	.522	S	.133	S	.133	S
2006	.193	S	.089	D	.333	S	.029	S	.543	S	.134	S	.127	S
2007	.194	S	.088	D	.315	D	.030	D	.564	S	.134	S	.121	S
2008	.195	S	.088	D	.298	D	.030	D	.585	S	.134	S	.116	S
2009	.196	S	.088	D	.280	D	.031	D	.606	D	.134	S	.110	S
2010	.197	S	.088	D	.263	D	.032	D	.627	D	.134	S	.104	S

PR (State Legislature Professionalism) 0 = least professionalized; 1 = most professionalized
GPC (Partisan Control of State Government) D = Democrat; S = Split/Divided; R = Republican

Source: Data compiled by author from various sources; see appendix A.

Appendix E

Georgia Presidential Election Results by County (2000–2012)

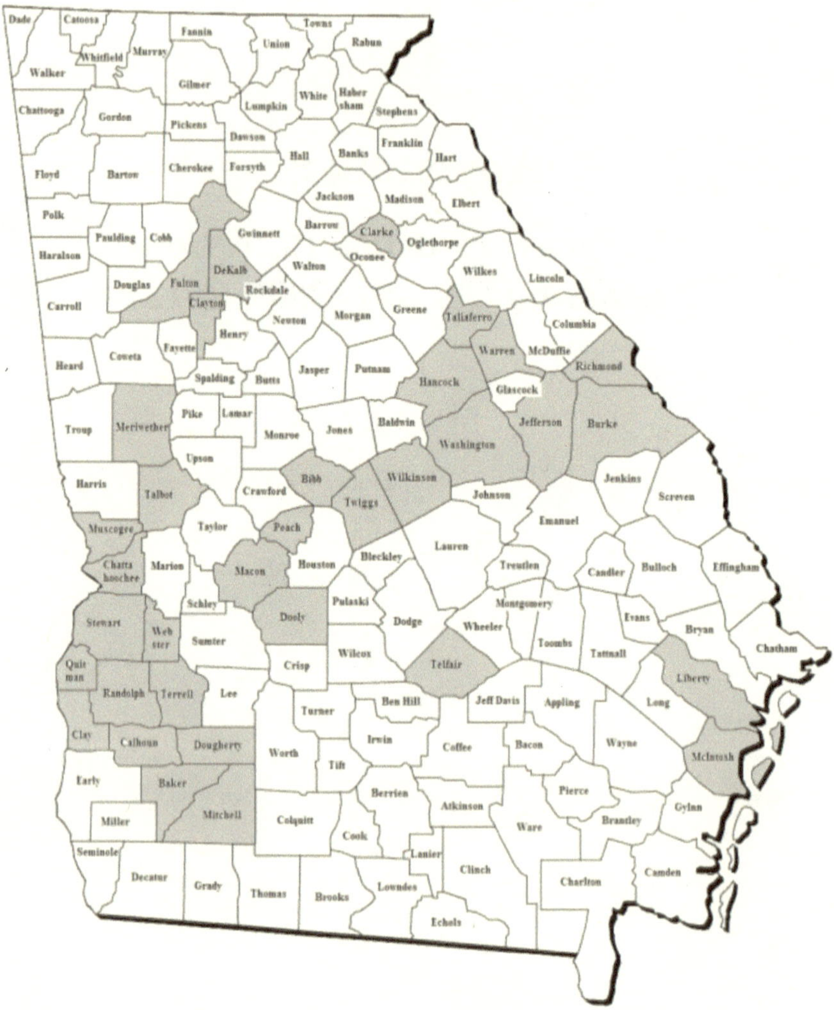

Figure E.1. Georgia Presidential Election Results 2000. Gray counties voted Democratic; white counties voted Republican.

Created by the author with data from the Georgia Secretary of State (n.d.).

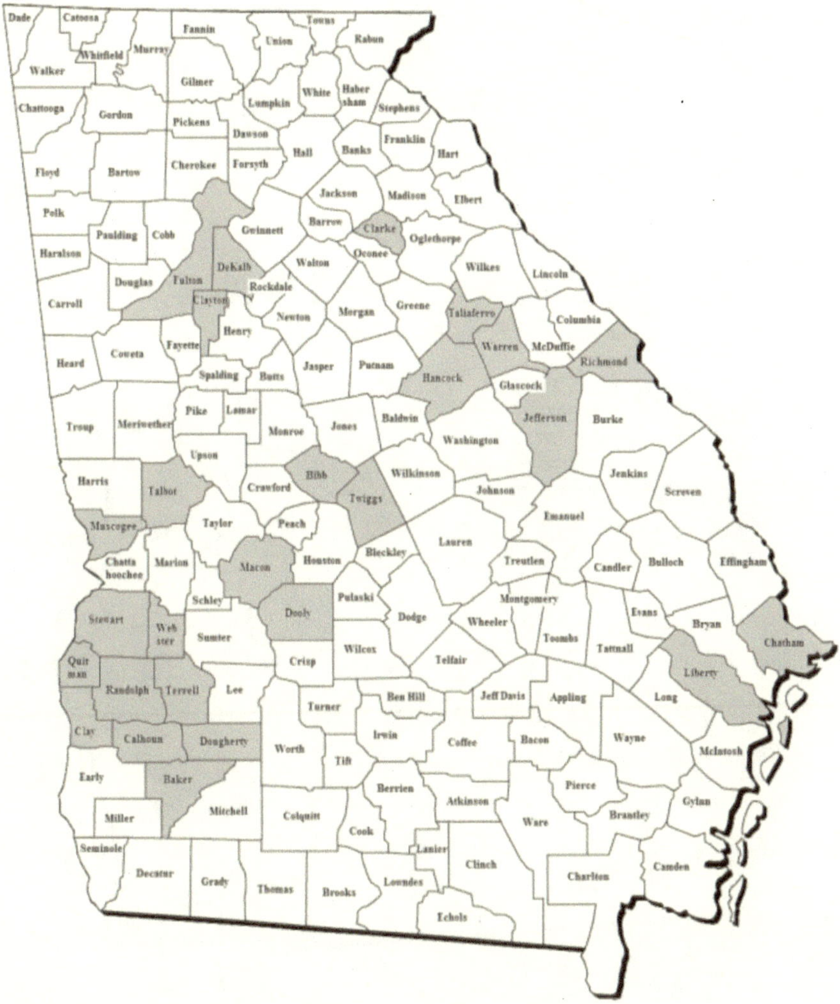

Figure E.2. Georgia Presidential Election Results 2004. Gray counties voted Democratic; white counties voted Republican.

Created by the author with data from the Georgia Secretary of State (n.d.).

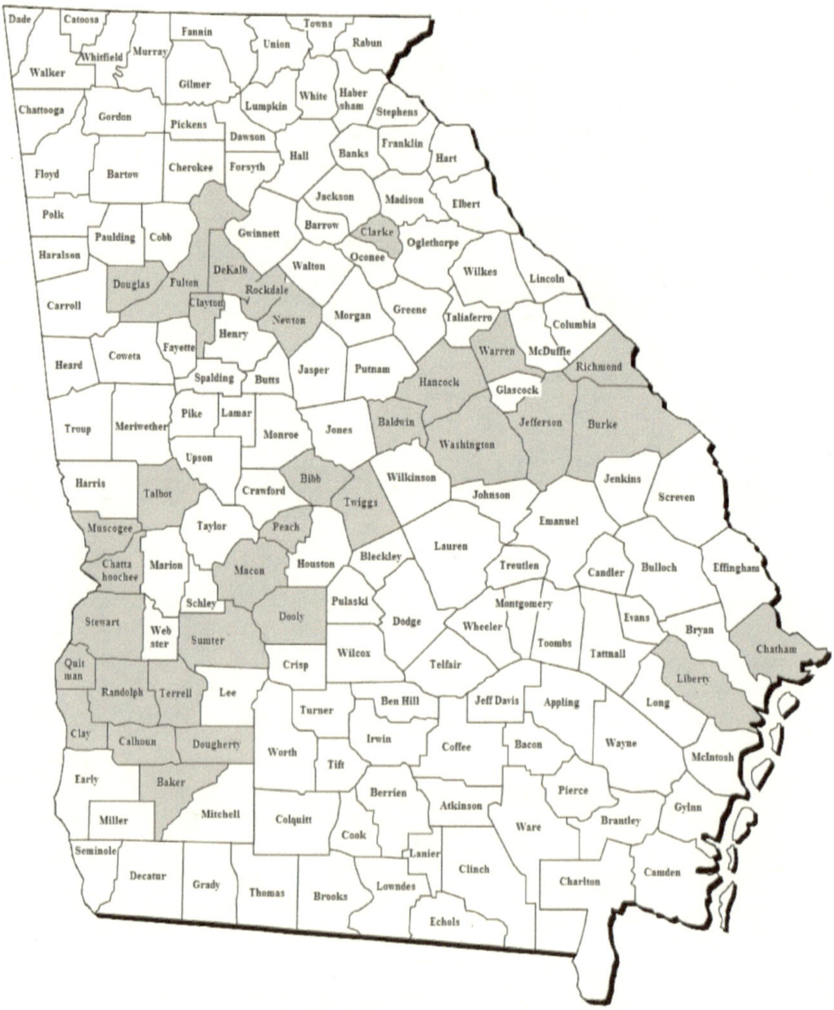

Figure E.3. Georgia Presidential Election Results 2008. Gray counties voted Democratic; white counties voted Republican.

Created by the author with data from the Georgia Secretary of State (n.d.).

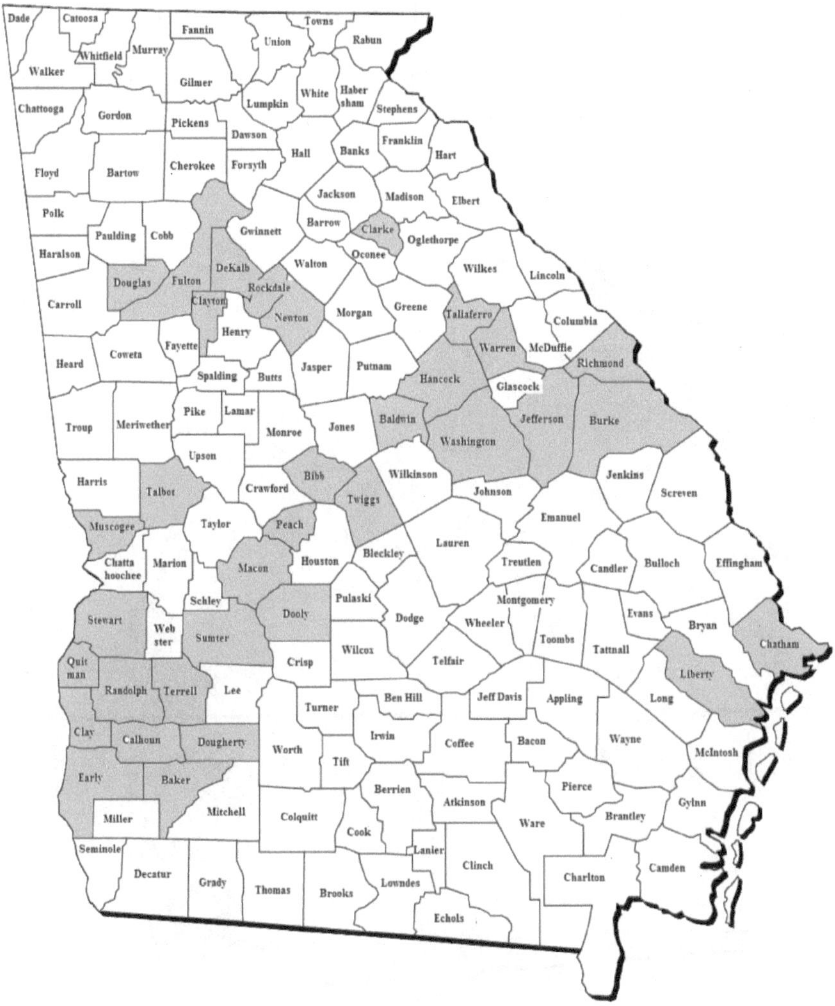

Figure E.4. Georgia Presidential Election Results 2012. Gray counties voted Democratic; white counties voted Republican.

Created by the author with data from the Georgia Secretary of State (n.d.).

GEORGIA POPULATION AND INCOME

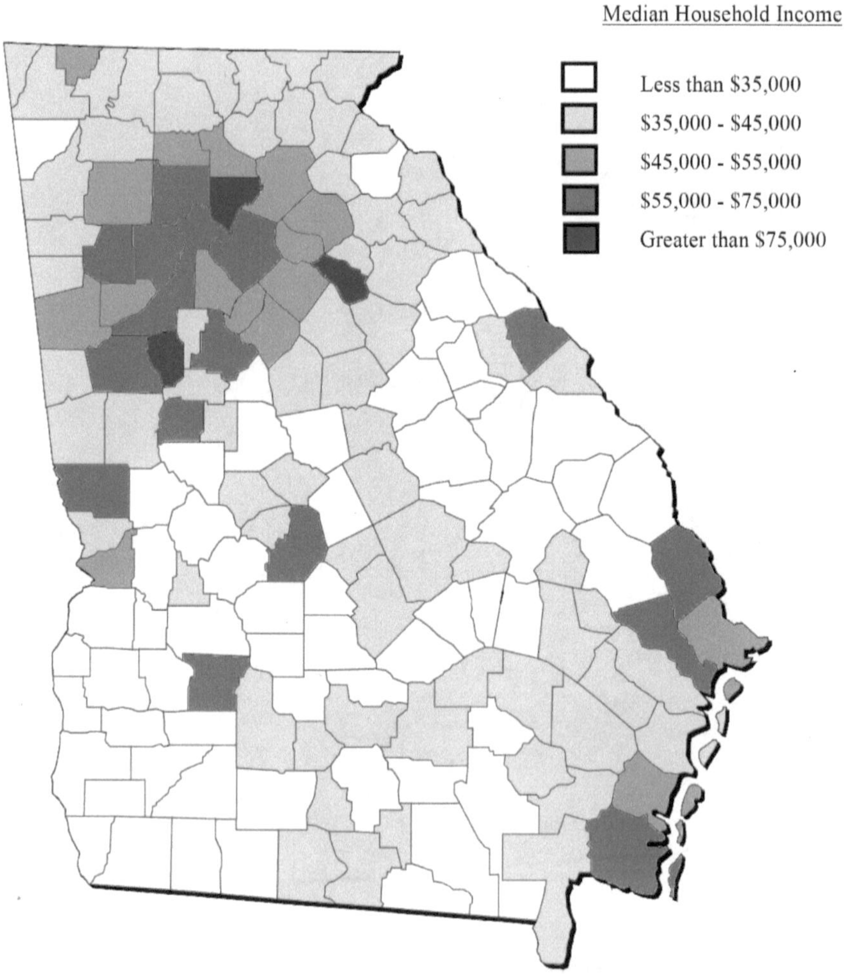

Figure E.5. Georgia Wealth Distribution

Created by the author with information obtained from the U.S. Census 2010.

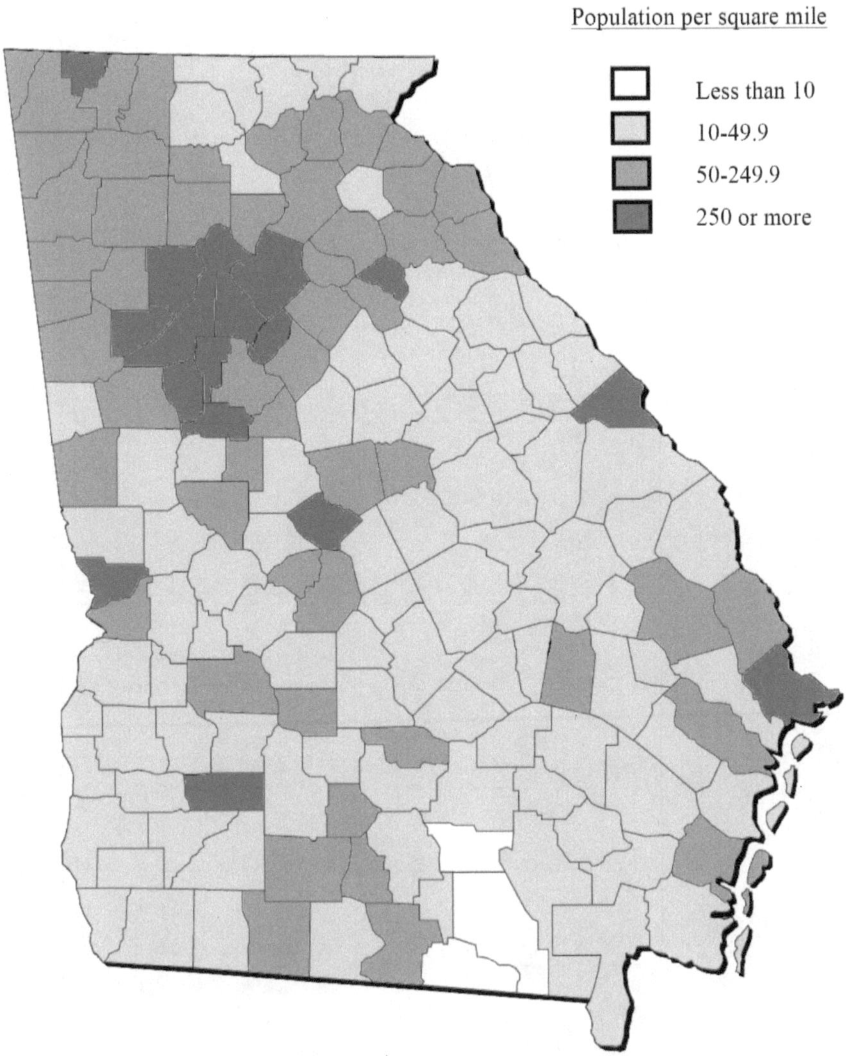

Figure E.6. Georgia Population Distribution
Created by the author with information obtained from the U.S. Census 2010.

Table E.1. Southeast Region Political Demographics

State Legislature Professionalism Ranking (PR), State Government Partisan Control (GPC), and Political Culture (PC)

Year	Alabama (PC 8.57)		Florida (PC 7.80)		Georgia (PC 8.80)		Kentucky (PC 7.40)		Mississippi (PC 9.0)		North Carolina (8.50)		South Carolina (8.75)		Tennessee (8.50)	
	PR	GPC	PR	GPC	PR	GPC	PR	GPC	PR	GPC	PR	GPC	PR	GPC	PR	GPC
1990	.122	S	.253	S	.123	D	.095	D	.147	D	.181	S	.181	S	.128	D
1991	.113	S	.252	D	.120	D	.094	D	.144	D	.176	S	.174	S	.126	D
1992	.103	S	.251	D	.117	D	.093	D	.140	S	.171	S	.166	S	.124	D
1993	.094	S	.251	S	.115	D	.091	D	.137	S	.165	D	.158	S	.122	D
1994	.085	S	.250	S	.112	D	.090	D	.134	S	.160	D	.150	S	.121	D
1995	.076	S	.250	S	.110	D	.088	D	.130	S	.154	S	.143	S	.119	S
1996	.067	S	.249	S	.107	D	.087	D	.127	S	.149	S	.135	S	.117	S
1997	.076	S	.245	S	.108	D	.096	D	.124	S	.156	S	.133	S	.117	S
1998	.085	S	.242	S	.110	D	.104	D	.121	S	.163	S	.132	S	.117	S
1999	.094	D	.238	R	.111	D	.113	D	.118	S	.170	S	.130	S	.117	S
2000	.104	D	.234	R	.112	D	.122	S	.116	D	.177	D	.129	S	.116	S
2001	.113	D	.230	R	.113	D	.131	S	.113	D	.184	D	.127	S	.116	S
2002	.122	D	.227	R	.115	D	.139	S	.110	D	.191	D	.126	S	.116	S
2003	.131	S	.223	R	.116	S	.148	S	.107	D	.198	S	.124	R	.116	D
2004	.122	S	.221	R	.116	S	.146	S	.108	S	.195	S	.130	R	.116	D
2005	.113	S	.217	R	.116	R	.144	S	.110	S	.192	D	.136	R	.117	S
2006	.105	S	.213	R	.116	R	.143	S	.111	S	.189	D	.143	R	.117	S
2007	.096	S	.210	R	.116	R	.141	S	.112	S	.186	D	.149	R	.117	S
2008	.087	S	.206	R	.116	R	.139	S	.114	S	.183	D	.155	R	.118	S
2009	.078	S	.210	R	.116	R	.137	S	.115	S	.180	D	.161	R	.118	S
2010	.069	S	.208	R	.116	R	.135	S	.116	S	.177	D	.167	R	.118	S

PR (State Legislature Professionalism) 0 = least professionalized; 1 = most professionalized
GPC (Partisan Control of State Government) D = Democrat; S = Split/Divided; R = Republican

Source: Data compiled by author from various sources; see appendix A.

Appendix F

State Average Professionalism Score and Ranking

Table F.1. State Average Professionalism Score and Ranking

Rank	*State*	*Squire Avg. Professionalism Score (1976–2010)*	*Rank*	*State*	*Squire Avg. Professionalism Score (1976–2010)*
1	California	0.5836	26	Oregon	0.1788
2	New York	0.5330	27	Louisiana	0.1547
3	Michigan	0.4982	28	Delaware	0.1523
4	Massachusetts	0.4150	29	Nevada	0.1462
5	Pennsylvania	0.3438	30	Virginia	0.1429
6	Wisconsin	0.3373	31	Mississippi	0.1408
7	Ohio	0.3317	32	Rhode Island	0.1340
8	Illinois	0.2830	33	Idaho	0.1329
9	Alaska	0.2654	34	Vermont	0.1306
10	Hawaii	0.2526	35	Indiana	0.1292
11	New Jersey	0.2490	36	Kansas	0.1277
12	Arizona	0.2365	37	Tennessee	0.1270
13	Florida	0.2338	38	West Virginia	0.1269
14	Colorado	0.2333	39	Maine	0.1228
15	Missouri	0.2273	40	Georgia	0.1226
16	Oklahoma	0.2132	41	Arkansas	0.1072
17	Maryland	0.2125	42	Kentucky	0.1064
18	Washington	0.2090	43	Alabama	0.1053
19	Texas	0.2050	44	New Mexico	0.0974
20	Iowa	0.1997	45	Montana	0.0900
21	Connecticut	0.1995	46	South Dakota	0.0771
22	Minnesota	0.1855	47	Utah	0.0737
23	South Carolina	0.1836	48	North Dakota	0.0633
24	North Carolina	0.1830	49	Wyoming	0.0600
25	Nebraska	0.1811	50	New Hampshire	0.0405
			National Average—.1977		

Source: Compiled by the author with data from Squire 2007.

References

Adams, Damian C. and Matthew J. Salois. 2010. "Local Versus Organic: A Turn in Consumer Preferences and Willingness-to-Pay." *Renewable Agriculture and Food Systems*, 25(4): 331–41.

Agricultural Issues Center. n.d. "Estimating California's Agricultural Exports." University of California. Accessed February 25, 2014. http://aic.ucdavis.edu/pub/exports.html.

Agricultural Marketing Service (AMS). n.d. "NOSB Aquaculture Working Group Interim Final Report." Accessed February 25, 2014. http://www.ams.usda.gov/AMSv1.0/getfile?dDocName=STELDEV3025623.

Agricultural Marketing Service (AMS). 2013. "Protecting Organic Crops from Synthetic Fertilizers." *Organic Integrity Quarterly*, April. Accessed April 5, 2013. http://www.ers.usda.gov/data-products/organic-production.aspx#.Uw6PsvldWSp.

Agricultural Marketing Service (AMS). 2016. "USDA Issues Proposed Rule to Amend Organic Livestock and Poultry Practices, Including Poultry Living Conditions; Invites Comments." Accessed December 15, 2016. https://www.ams.usda.gov/press-release/usda-issues-proposed-rule-amend-organic-livestock-and-poultry-practices-including.

Agriculture Specialist 1. Interviewed by Samantha Mosier. Written communication, October 24, 2013.

Agriculture Specialist 2. Interviewed by Samantha Mosier. Tape recording/transcription, October 27, 2013.

Allen, Mahalley D., Carrie Pettus, and Donald P. Haider-Markel. 2004. "Making the National Local: Specifying the Conditions for National Government Influence on State Policymaking." *State Politics and Policy Quarterly*, 4(3): 318–44.

Alston, Julian M. and Philip G. Pardey. 2008. "Public Funding for Research into Specialty Crops." *HortScience*, 43(5): 1461–70.

Altieri, Miguel A. 1990. "Why Study Traditional Agriculture?" In *The Ecology of Agricultural Systems*. Edited by R. Carroll, J. H. Van der Meer, and P. M. Rossett. New York: MacGraw Hill Publ., 551–64.

Assembly Republican Committee. 2002. April 17 Assembly Republican Bill Analysis for Agriculture Committee. Senator Representative Polanco Files, AB 2823. California State Archives, Sacramento, CA.

Assembly Republican Committee. 2005. April 19 Assembly Republican Bill Analysis for Agriculture Committee. Senate Republican Committee, SB 730. California State Archives, Sacramento, CA.

Associated Press. "Butz Caught in Uproar for Racial Slurs." Rome News-Tribune, October 1976.

Bailey, Michael A. and Mark Carl Rom. 2004. "A Wider Race? Interstate Competition Across Health and Welfare Programs." *Journal of Politics*, 66(2): 326–47.

Baker, Brian. 2005. "Brief History of Organic Farming and the National Organic Program." Organic Materials Review Institute. Accessed February 25, 2014. http://www.sarep.ucdavis.edu/sfr/organic/files-images/intro2.pdf.

Ball, Eldon, Sun Ling Wang, and Richard Nehring. 2013. "Agricultural Productivity in the U.S." Economic Research Service, United States Department of Agriculture. Accessed December 22, 2013. http://www.ers.usda.gov/data-products/agricultural-productivity-in-the-us.aspx#.UrczcfRDuSo.

Balla, Steven J. 2001. "Interstate Professional Associations and the Diffusion of Policy Innovations." *American Politics Research*, 29(3): 221–45.

Barham, James, Debra Tropp, Kathleen Enterline, Jeff Farbman, John Fisk, and Stacia Kiraly. 2012. *Regional Food Hub Resource Guide*. US Department of Agriculture, Agricultural Marketing Service. Access January 31, 2016. https://www.ams.usda.gov/sites/default/files/media/Regional%20Food%20Hub%20Resource%20Guide.pdf.

Baumgartner, Frank R. and Bryan D. Jones. 2009. *Agendas and Instability in American Politics*. 2nd ed. Chicago: University of Chicago Press.

Belasco, Warren James. 2006. *Appetite for Change: How the Counterculture Took On the Food Industry*. 2nd ed. Ithaca, NY: Cornell University Press.

Berry, Frances Stokes and William D. Berry. 1990. "State Lottery Adoptions as Policy Innovations: An Event History Analysis." *The American Political Science Review*, 84(2): 395–415.

Berry, Frances Stokes and William D. Berry. 2007. "Innovation and Diffusion Models in Policy Research." In *Theories of the Policy Process*. Edited by Paul A. Sabatier. Boulder, CO: Westview Press, 223–60.

Berry, William D. and Brady Baybeck. 2005. "Using Geographic Information Systems to Study Interstate Competition." *The American Political Science Review*, 99(4): 505–19.

Bezio, Meghan. 2009. "An Agricultural History of Manchester, Vermont." University of Vermont Barn Census History. Accessed October 28, 2015. http://www.uvm.edu/~hp206/2009/Manchester/History.html.

Bidwell, Percy Wells and John I Falconer.1941 *History of Agriculture in the Northern United States*. New York: Carnegie Institution of Washington.

Birkland, Thomas. 1998. "Focusing Events, Mobilization, and Agenda Setting." *Journal of Public Policy*, 18(1): 53–74.

Blobaum, Roger. 2010. "A Speech 'The Evolution of Organic Research from the 1970s to 2010' Given at the University of Minnesota Lamberton Research Station." Accessed April 4, 2013. http://rogerblobaum.com/the-evolution-of-organic-research-from-1970s-to-2010/.

Boeckelman, Keith. 1992. "The Influence of States on Federal Policy Adoptions." *Policy Studies Journal*, 20(3): 365–75.

Boehmke, Frederick J. and Paul Skinner. 2012. "State Policy Innovativeness Revisited." *State Politics and Policy Quarterly*, 12(3): 304–30.

Boehmke, Frederick J. and Richard Witmer. 2004. "Disentangling Diffusion: The Effects of Social Learning and Economic Competition on State Policy Innovation and Expansion." *Political Research Quarterly*, 57(1): 39–51.

Bones, Gordon G. 1992. "State and Federal Organic Food Certification Laws: Coming of Age." *North Dakota Law Review*, 68: 405–44.

Bonner, James C. 1964. *A History of Georgia Agriculture 1732–1860*, 2009 Paperback Edition. Athens: University of Georgia Press.

Bonti-Ankomah, Samual and Emmanuel K Yiridoe. 2006. *Organic and Conventional Food: A Literature Review of the Economics of Consumer Perceptions and Preferences*. Organic Agriculture Centre of Canada.

Boushey, Graeme. 2010. *Policy Diffusion Dynamics in America*. New York: Cambridge University Press.

Brians, Janet, and Grant Brians. Interviewed by Ellen Farmer. Cultivating a Movement: Organic and Sustainable Farming, Regional History Project. University Library, University of California-Santa Cruz, July 19, 2007.

Bricmont, Barney. Interviewed by Ellen Farmer. Cultivating a Movement: Organic and Sustainable Farming, Regional History Project. University Library, University of California, Santa Cruz, March 7, 2007.

Brooks, Sarah M. 2005. "Interdependent and Domestic Foundations of Policy Change: The Diffusion of Pension Privatization around the World." *International Studies Quarterly* 49(2): 273–94.

Brown, Louie A. 2002a. A Letter to Honorable Darrell Steinberg on May 6. Legislative Council Digest, AB 2823, California State Archives, Sacramento, CA.

Brown, Harold R. 2002b. *The Greening of Georgia: The Improvement of the Environment of Twentieth Century*. Macon: Mercer University Press.

Brush, Marc and Connor Link. 2012. "How to Make Organic Relevant in 2013." *Nutrition Business Journal* presentation. Accessed April 4, 2013. http://newhope360.com/research-and-insights/how-make-organic-relevant-2013#slide-3-field_images-280291.

Bureau of Economic Analysis (BEA). 2015a. "Vermont BEARFACTS." Accessed October 25, 2015. https://www.bea.gov/regional/bearfacts/pdf.cfm?fips=50000&areatype=STATE&geotype=3.

Bureau of Economic Analysis (BEA). 2015b. "Northeast Regional Data GDP and Personal Income." Accessed October 25, 2015. https://www.bea.gov/itable/iTable.cfm?ReqID=70&step=1#reqid=70&step=10&isuri=1&7003=200&7035=-1&7004=naics&7005=-1&7006=50000&7036=1&7001=1200&7002=1&7090=70&7007=2014,2013,2012,2011,2010,2009,2008&7093=levels.

Burroughs, Kate. 2004. "Why Organic?" *California Certified Organic Magazine*, XXI(1).

Butz, Earl.1971. "Crisis or Challenge?" *Nation's Agriculture*, 46(6): 19.

California Cattlemen's Association. 1989. Letter to the Honorable Delaine Eastin. Assembly Committee on Government Efficiency and Consumer Protection, AB 2012. California State Archives, Sacramento, CA.

California Certified Organic Farmers (CCOF). 1988. "Our Story So Far: CCOF Marks 15th Anniversary." *California Certified Organic Farmers Statewide Newsletter*, 5(2): 1–3. Accessed February 25, 2014. http://www.ccof.org/sites/default/files/media/Documents/history_ourstorysofar1988.pdf.

California Certified Organic Farmers (CCOF). 2013a. "Our History." Accessed February 25, 2014. http://www.ccof.org/ccof.

California Certified Organic Farmers (CCOF). 2013b. "Who is CCOF?" Accessed February 25, 2014. http://www.ccof.org/ccof.

California Department of Finance. 1990. AB 2012 Bill Analysis. Senate Appropriations Committee, AB 2012. California State Archives, Sacramento, CA.

California Department of Health Services (CDHS). 1979. Bill Analysis of AB443. Assemblyman Gage Papers, AB443. California State Archives, Sacramento, CA.

California Department of Health Services (CDHS). 1982. Bill Analysis of AB 3422. Assemblyman Sam Farr Papers, AB 3422. California State Archives, Sacramento, CA.

California Department of Health Services (CDHS). 1989. Letter to the Honorable Delaine Eastin. Assembly Committee on Government Efficiency and Consumer Protection, AB 2012. California State Archives, Sacramento, CA.

California Grocers Association. 1989. Letter to Barry Epstein. Assembly Committee on Governmental Efficiency and Consumer Protection File, AB2012. California State Archives, Sacramento, CA.

California Secretary of State. n.d. Prior Elections Statewide Election Results Database. http://www.sos.ca.gov/elections/prior-elections/statewide-election-results/

Canon, Bradley C. and Lawrence Baum. 1981. "Patterns of Adoption of Tort Law Innovations: An Application of Diffusion Theory to Judicial Doctrines." *The American Political Science Review*, 75(4): 975–87.

Cantisano, Amigo Bob. 1982. Letter to fellow ecological advocates, farmers, and consumer alike. Assemblyman Sam Farr Papers, AB 3422. California State Archives, Sacramento, CA.

Cantisano, Amigo Bob. Interviewed by Sarah Rabkin. Cultivating a Movement: Organic and Sustainable Farming, Regional History Project. University Library, University of California, Santa Cruz, April 7 and 9, 2008.

Carolina Farm Stewards (CFSA). 2014. "What We Do." Accessed January 20, 2014. http://www.carolinafarmstewards.org/whatwedo/.

Center for Agroecology and Sustainable Food Systems (CASFS). 2010a. "Farm Garden Projects." University of California-Santa Cruz. Accessed January 15, 2014. http://casfs.ucsc.edu/about/history/farm-garden-projects.

Center for Agroecology and Sustainable Food Systems (CASFS). 2010b. "Facilities." University of California-Santa Cruz. Accessed January 15, 2014. http://casfs.ucsc.edu/about/facilities.

Charles, Dan. 2016. "Congress Just Passed a GMO Labeling Bill. Nobody's Super Happy About It." *NPR.* Accessed December 16, 2016. http://www.npr.org/sections/thesalt/2016/07/14/486060866/congress-just-passed-a-gmo-labeling-bill-nobodys-super-happy-about-it.

Chez Panisse. 2012. "About Chez Panisse." Accessed February 25, 2014. http://www.chezpanisse.com/about/chez-panisse/.

Christensen, Andy. 1990. April 12 Testimony to Vermont House of Representatives. Obtained Via Personal Communications with Enid Wonnacott November 2, 2015.

Christensen, Clair, Deborah Giraud, Refugio Gonzalez, Liese Greensfelder, Pedro Ilic, Manuel Jimenez, Steve Mendivil, Faustino Munoz, Claudia Myers, John Stumbos, Louie Valenzuela, Ron Voss, and Paul Vossen. 1990. "The UC Small Farm Program." UC Small Farm Program, University of California-Davis. Accessed February 25, 2014. http://sfp.ucdavis.edu/files/143900.pdf.

Clark, Jill. 1985. "Policy Diffusion and Program Scope: Research Directions." *Publius*, 15(4): 61–70.

Clark, Lisa F. 2015. *The Changing Politics of Organic Food in North America.* Northhampton, MA: Edward Elgar.

Cobb, Roger W. and Charles D. Elder. 1983. *Participation in American Politics: The Dynamics of Agenda Building*. Baltimore, MD: John Hopkins University Press.

Cohen, Micah. 2012. "'New' Vermont is Liberal, but 'Old' Vermont is Still There." *The New York Times*. Accessed October 26, 2015. http://fivethirtyeight.blogs.nytimes.com/2012/10/01/new-vermont-is-liberal-but-old-vermont-is-still-there/?_r=0.

Cohen-Vogel, Lora and William Kyle Ingle. 2007. "When Neighbours Matter Most: Innovation, Diffusion and State Policy Adoption in Tertiary Education." *Journal of Education Policy,* 22(3): 241–62.

Community Alliance with Family Farmers (CAFF). 2013. "About." Accessed February 25, 2014. http://caff.org/about/.

Concerned Citizen 1. 2002. Letter to Assemblymember Strom-Martin on August 16. Virginia Strom-Martin Papers, AB 2823. California State Archives, Sacramento, CA.

Concerned Citizen 2. 2002. Letter to Assemblymember Strom-Martin on August 16. Virginia Strom-Martin Papers, AB 2823. California State Archives, Sacramento, CA.

Concerned Citizen 3. 2002. Letter to Assemblymember Strom-Martin on August 17. Virginia Strom-Martin Papers, AB 2823. California State Archives, Sacramento, CA.

Cone, Marla. 2010. "Insecticide to Be Banned—Three Decades After Tainted Melons Sickened 2,000 People." Environmental Health News. Accessed February 25, 2014. http://www.environmentalhealthnews.org/ehs/news/aldicarb-phaseout.

Costanigro, Marco, Stephen Kroll, Dawn Thilmany and Marisa Bunning. 2014. "Is it Love for Local/Organic or Hate for Conventional? Asymmetric Effects of Information and Taste on Label Preferences in an Experimental Auction." Food Quality and Preferences, 31(January): 94–105.

Daley, Dorothy M. and James C. Garand. 2005. "Horizontal Diffusion, Vertical Diffusion, and Internal Pressure in State Environmental Policymaking, 1989–1998." *American Politics Research*, 33(5): 615–44.

Davies, Anne, Albert J. Titterington, and Clive Cochrane. 1995. "Who Buys Organic Food? A Profile of the Purchasers of Organic Food in Northern Ireland." *British Food Journal*, 97(10): 17–23.

Dettmann, Rachel L. and Carolyn Dimitri. 2007. "Organic Consumers: A Demographical Portrayal of Organic Vegetable Consumption within the United States." Paper presented at the 105th EAAE Seminar.

DiMatteo and Gershuny. 2007. "The Organic Trade Association." In *Organic Farming: An International History*. Edited by William Lockeretz. Cambridge: CABI International.

Duffy, Robert J. 2003. *The Green Agenda in American Politics: New Strategies for the Twenty-First Century*. Lawrence, KS: University Press of Kansas.

Economic Research Service (ERS). 2009. "Emerging Issues in the U.S. Organic Industry." United States Department of Agriculture, Economic Information Bulletin Number 55.

Economic Research Service (ERS). 2013a. "Organic Production State-Level Tables, Table 4." Accessed February 25, 2014. http://www.ers.usda.gov/data-products/organic-production.aspx#.Uw6PsvldWSp.

Economic Research Service (ERS). 2013b. "State Ranking Tables, Table 20." Accessed February 25, 2014. http://www.ers.usda.gov/data-products/agricultural-productivity-in-the-us.aspx#28270.

Economic Research Service (ERS). 2013c. "State Ranking Tables, Table 22." Accessed February 25, 2014. http://www.ers.usda.gov/data-products/agricultural-productivity-in-the-us.aspx#28270.

Egide, Mark A. 2002. Letter to Assembly Member Strom-Martin from Avalon Natural Products on August 16. Virginia Strom-Martin Papers, AB 2823. California State Archives, Sacramento, CA.

Elazar, Daniel. 1972. *American Federalism: A View from the States*. 2nd Ed. New York: Thomas Y. Crowell Company.

Elazar, Daniel. 1994. *The American Mosaic: The Impact of Space, Time, and Culture on American Politics*. Boulder, CO: Westview Press.

Employment Development Department (EDD). 2010. "California Agricultural Bulletin 3rd Quarter." State of California. Accessed February 25, 2014. http://www.calmis.ca.gov/file/agric/2010–3cab.pdf.

Epstein, Barry. 1989a. Letter to Ann Evans and Mike Falasco. Assembly Committee on Governmental Efficiency and Consumer Protection File, AB2012. California State Archives, Sacramento, CA.

Epstein, Barry. 1989b. CCOF Position on the Organic Foods Labeling and Enforcement Act of 1989. Senate Committee on Health and Human Services, AB2012. California State Archives, Sacramento, CA.

Epstein, Barry. Interviewed by Samantha Mosier. Tape recording/transcription. December 23, 2013.

Evans, Ann (Chairwoman). 1989. Assembly Committee on Governmental Efficiency and Consumer Protection Hearing May 10, 1989. Assembly Committee on Governmental Efficiency and Consumer Protection File, AB2012. California State Archives, Sacramento, CA.

Fairbanks, David. 1979. "Politics, Economics, and the Public Morality: Why Some States Are More 'Moral' Than Others." *Policy Studies Journal*, 7(4): 714–21.

Fagerberg, Jan. 2005. "Innovation: A Guide to the Literature." In *The Oxford Handbook of Innovation*. Edited by Jan Fagerberg, David C. Mowery, and Richard R. Nelson. New York: Oxford University Press, 1–26.

Farr, Sam. Interviewed by Ellen Farmer. Cultivating a Movement: Organic and Sustainable Farming, Regional History Project. University Library, University of California, Santa Cruz, August 23, 2007.

Farr, Sam. Interviewed by Samantha Mosier. Tape recording/transcription. December 12, 2013.

Federal Trade Commission. 1979. "Annual Report of Federal Trade Commission." Accessed February 25, 2014. http://www.ftc.gov/sites/default/files/documents/reports_annual/annual-report-1979/ar1979_0.pdf.

Federal Trade Commission. 1980. "Annual Report of Federal Trade Commission, 1980." Accessed February 25, 2014. https://www.ftc.gov/sites/default/files/documents/reports_annual/annual-report-1980/ar1980_0.pdf.

FiBL and IFOAM. 2013. *The World of Organic Agriculture*. The Research Institute of Organic Agriculture.

FiBL and IFOAM. 2016. *The World of Organic Agriculture*. The Research Institute of Organic Agriculture (FibL). Accessed November 1, 2016. https://shop.fibl.org/fileadmin/documents/shop/1698-organic-world-2016.pdf.

Fishman, Stuart. 1979a. Notes on Meeting with Assemblyman Gage February 6. Assemblyman Gage Papers, AB443. California State Archives, Sacramento, CA.

Fishman, Stuart. 1979b. Some Plain Talk about the Organic Food Bill: AB 443. Assemblyman Gage Papers, AB443. California State Archives, Sacramento, CA.

Fishman, Stuart. 1979c. Form Letter to be Sent Out to John Garamendi. Assemblyman Gage Papers, AB443. California State Archives, Sacramento, CA.

Fishman, Stuart. 1979d. Notes on Meeting with Farmers March 1979. Assemblyman Gage Papers, AB443. California State Archives, Sacramento, CA.

Fishman and Fishman. 1979. Comments Received by Gage's Office. Assemblyman Gage Papers, AB443. California State Archives, Sacramento, CA.

Flatt, William P. 2013. "Agriculture in Georgia: Overview." *New Georgia Encyclopedia*. Accessed January 1, 2014. http://www.georgiaencyclopedia.org/articles/business-economy/agriculture-georgia-overview.

Food and Agriculture Organization of the United Nations (FAO). n.d. "Organic Agriculture FAQ." Accessed February 25, 2014. http://www.fao.org/organicag/oa-faq/oa-faq3/en/.

Former GDA (Georgia Department of Agriculture) Staff Member 1. Interviewed by Samantha Mosier. Tape recording/transcription. October 19, 2013.

Former GDA (Georgia Department of Agriculture) Staff Member 2. Interviewed by Samantha Mosier. Written communication. November 4, 2013.

Former NOFA-VT Staff Member. Interviewed by Samantha Mosier. Tape recording/transcription. October 14, 2015.

Former Organic Food Advocate. Interviewed by Samantha Mosier. Tape recording/transcription. October 17, 2013.

Former VOF Staff Member. Interviewed by Samantha Mosier. Tape recording/tran scription. October 2, 2015.

Freeman, Patricia K. 1985. "Interstates Communication among State Legislators Regarding Energy Policy Innovation." *Publius*, 15(4): 99–111.

Gabriel, Doreen, Stephen J. Carver, Helen Durham, William E. Kunin, Robert C. Palmer, S. M. Sait, S. Stagl, and Tim G. Benton. 2009. "The Spatial Aggregation of Organic Farming in England and its Underlying Environmental Correlates." *Journal of Applied Ecology*, 46(2): 323–33.

Georgia Crop Improvement Association Organic Certification Program. 2013. "Administration and Policy Manual." Accessed February 25, 2014. http://www.certi fiedseed.org/PDF/UGAHosted/OrganicAdmin.pdf.

GDA (Georgia Department of Agriculture) Staff Member. Interviewed by Samantha Mosier. Written communication. October 29, 2013.

Georgia Farm Bureau. 2014. "Georgia Agriculture—The State's $71 Billion Industry." Accessed February 25, 2014. http://www.gfb.org/aboutus/georgia_agriculture.html.

Georgia Organics. 2012. "About Us." Accessed February 25, 2014. http://georgia organics.org/about-us/overview.

Georgia Secretary of State. n.d. "Current and Past Election Results." http://sos .ga.gov/index.php/Elections/current_and_past_elections_results.

Gershuny, Grace. 1986. "A Brief History of NOFA VT." Personal Communications with Enid Wonncott 2015.

Gershuny, Grace. 1990 (January 24). Testimony Re: S.277 for the Vermont Senate Agriculture Committee. State of Vermont Archives A116–00381.

Gershuny, Grace. 2002. "The Organic Revolution." Institute for Social Ecology. Archived website accessed February 25, 2014. http://www.social-ecology.org/author /grace-gershuny/.

Gershuny, Grace. 2016. *Organic Revolutionary: A Memoir of the Movement for Real Food, Planetary Healing, and Human Liberation*. Joes Book Press.

Glick, Henry R. and Scott P. Hays. 1991. "Innovation and Reinvention in State Policymaking: Theory and the Evolution of Living Will Laws." *The Journal of Politics*, 53(3): 835–50.

Glyn, Noah and Scott Drenkard. 2013. "Prop 13 in California, 35 Years Later." Tax Foundation. Accessed February 25, 2014. http://taxfoundation.org/blog/prop-13 -california-35-years-later.

Goldstein, Richard. 2008. "Earl L. Butz, Secretary Felled by Racial Remark, is Dead at 98." *New York Times*. February 4. Accessed April 2, 2013. http://www.nytimes .com/2008/02/04/washington/04butz.html?_r=0.

Graham, Erin R, Charles R Shipan, and Craig Volden. 2013. "The Diffusion of Policy Diffusion Research in Political Science." *British Journal of Political Science*, 43(3): 673–701.

Graham, Erin R., Charles R. Shipan, and Craig Volden. 2014. "The Communication of Ideas across Subfields in Political Science." *PS: Political Science and Politics*, 47(2): 468–76.

Gray, Charles M., Virginia Gray, and Bruce Williams. 1981. "Federalism, Policy, and Innovation in Corrections." *Policy Studies Review*, 1(2): 288–329.

Gray, Virginia. 1973a. "Innovation in the States: A Diffusion Study." *The American Political Science Review*, 67(4): 1174–85.

Gray, Virginia. 1973b. "Rejoinder to 'Comment' by Jack L. Walker." *The American Political Science Review*, 67(4): 1192–93.

Gray, Virginia. 2012. "The Socioeconomic and Political Context of States." In *Politics in the American States: A Comparative Analysis*. Edited by Virginia Gray, Russell Hanson, and Thad Kousser. San Diego, CA: University of California Press, 1–29.

Green, Ray. 2002. Johnson CC'ed email. Virginia Strom-Martin Papers, AB 2823. California State Archives, Sacramento, CA.

Greene, Catherine R. 2013. "Growth Patterns in the U.S. Organic Industry." Amber Waves. October. Accessed February 25, 2014. http://www.ers.usda.gov/amber-waves/2013-october/growth-patterns-in-the-us-organicindustry.aspx#.Uwz4-vldWSq.

Grupp, Fred W., Jr., and Alan R. Richards. 1975. "Variations in Elite Perceptions of American States as Referents for Public Policy Making." *The American Political Science Review*, 69(3): 850–58.

Guthman, Julie. 2004. *Agrarian Dreams: The Paradox of Organic Farming in California*. Berkeley: University of California Press.

Hagen, Kenneth S., William W. Allen, and Richard L. Tassan. 1981. "Mediterranean Fruit Fly: The Worst May Be Yet to Come." *California Agriculture*, March/April: 6–7.

Haider-Markel, Donald P. 2001. "Policy Diffusion as a Geographical Expansion of the Scope of Political Conflict: Same-Sex Marriage Bans in the 1990s." *State Politics and Policy Quarterly*, 1(1): 5–26.

Hancock, Dennis W., Glen H. Harris, Randy W. Franks, Steven P. Morgan, and T. Wade Greene. 2011. "Soil and Fertilizer Management Considerations for Forage Systems in Georgia." The University of Georgia Cooperative Extension. Accessed February 25, 2014. http://www.caes.uga.edu/applications/publications/files/pdf/B%201346_3.PDF.

Hays, Scott P. 1996. "Patterns of Reinvention: The Nature of Evolution During Policy Diffusion." *Policy Studies Journal*, 24(4): 551–66.

Haystead, Ladd, and Gilbert D Fite. 1955. *The Agricultural Regions of the United States*. Norman, OK: University of Oklahoma Press.

Hite, Mark. 1993. AB 1713 Hearing 7/13/93. Senate Committee on Agriculture and Water Resources, AB 1713. California State Archives, Sacramento, CA.

Hooker, Neal and Christopher J. Shanahan. 2012. "Emerging Spatial Dependencies with U.S. Organic Supply Chains." *Journal of Food Products Marketing*, 18(5): 426–50.

Hughner, Renée Shaw, Pierre McDonagh, Andrea Prothero, Clifford J. Schultz II, and Julie Stanton. 2007. "Who are Organic Food Consumers? A Compilation and Review of Why People Purchase Organic Food." *Journal of Consumer Behavior*, 6: 1–17.

Ingram, Mrill and Helen Ingram. 2005. "Creating Credible Edibles: The Organic Agriculture Movement and the Emergence of U.S. Federal Organic Standards." In *Routing the Opposition: Social Movements, Public Policy and Democracy*. Edited by David S. Meyer, Valerie Jenness, and Hell Ingram. Minneapolis: University of Minnesota Press, 121–48.

Jacobsson, Staffan and Volkar Lauber. 2006. "The Politics and Policy of Energy System Transformation—Explaining the German Diffusion of Renewable Energy Technology." *Energy Policy*, 34(3): 256–76.

Jarvis, William T. 1979. Letter to Assemblyman Michael Gage. Assemblyman Gage Papers, AB443. California State Archives. Sacramento, CA.

Jessup, Bill. n.d. Letter to Stuart (Fishman) 2/28. Assemblyman Gage Papers, AB443. California State Archives. Sacramento, CA.

Johnson, Renée. 2008. "Organic Agriculture in the United States: Program and Policy Issues." Congressional Research Service. Accessed March 3, 2013. http://assets.opencrs.com/rpts/RL31595_20081125.pdf.

Johnston, Warren E. 2003. "Cross Sections of a Diverse Agriculture: Profiles of California's Agricultural Production Regions and Principal Commodities." In *California Agriculture: Dimensions and Issues*. Edited by Jerry Siebert. Giannini Foundation of Agricultural Economics, University of California.

Jones, Walter C. 2010. "Tommy Irvin's Retirement Caps Long Political Career." *Times-Herald.* November 17. Accessed February 25, 2014. http://www.times-herald.com/opinion/Tommy-Irvins-retirement-caps-long-political-career-1377331.

Jordan, Andrew and Dave Huitema. 2014. "Innovations in Climate Policy: The Politics of Invention, Diffusion and Evaluation." *Environmental Politics,* 23(5): 715–34.

Kamieniecki, Sheldon. 1997. "Political Parties and Environmental Policy." In *Environmental Politics and Policy: Theories and Evidence*. Edited by James P. Lester. United States: Duke University Press, 146–67.

Karch, Andrew. 2006. "National Intervention and the Diffusion of Policy Innovations." *American Politics Research*, 34: 403–26.

Karch, Andrew. 2007. *Democratic Laboratories: Policy Diffusion Among the American States*. Ann Arbor: University of Michigan Press.

Kaufmanna, Peter, Romualdas Zemeckis, Virgilius Skulskis, Emilija Kairyte, and Sigrid Stagl. 2011. "The Diffusion of Organic Farming in Lithuania." *Journal of Sustainable Agriculture*, 35: 522–49.

Kerselaers, E., L. D. Cock, L. Lauwers, and G. V. Huylenbroeck. 2007. "Modelling farm level economic potential for conversion to organic farming." *Agricultural Systems* 94: 671–82.

Kingdon, John. 2003. *Agendas, Alternatives, and Public Policies*. 2nd ed. New York: Longman.

Kizer, Kenneth W. (signed by Stephen W. Kissler). 1989. Letter from CDHS to the Honorable Delaine Eastin. Assembly Committee on Government Efficiency and Consumer Protection, AB 2012. California State Archives, Sacramento, CA.

Klarner, Carl. 2013. "Klarner Politics, State Partisan Balance Data." Accessed May 15, 2013. http://www.indstate.edu/polisci/klarnerpolitics.htm.

Klonsky, Karen. 2000. "Forces Impacting the Production of Organic Foods." *Agriculture and Human Values*, 17(3): 233–43.

Klonsky, Karen. 2003. "Organic Agricultural Production in California." In *California Agriculture: Dimensions and Issue*. Edited by Jerry Siebert. Giannini Foundation of Agricultural Economics, University of California.

Knorr, Dietrich. 1982. "Natural and Organic Foods: Definitions, Quality, and Problems." *Cereal Foods World*, 27(4): 163–68.

Knorr, Dietrich. 1984. "Feasibility of Analytical Procedures and Unit Operations for the Distinction Between Organic, Natural or Conventional Foods." *Biological Agriculture and Horticulture*, 2(2): 183–94.

Kraft, Michael E. 2012. "Environmental Policy in Congress." In *Environmental Policy: New Directions for the Twenty-First Century*. Edited by Norman J. Vig and Michael E. Kraft. Washington, D.C.: CQ Press, 99–124.

Kwon, Myungjung, Frances Berry, and Richard Feiock. 2009. "Understanding the Adoption and Timing of Economic Development Strategies in U.S. Cities Using Innovation and Institutional Analysis." *Journal of Public Administration Research and Theory*, 19(4): 967–88.

Lambert, Gene. 2002. Letter to Michael F. Thompson from Covington and Burling on August 29. Virginia Strom-Martin Papers, AB 2823. California State Archives, Sacramento, CA.

Lang, Corey and Shanna Pearson-Merkowitz. 2013. "Partisan Sorting in the U.S. States, 1972–2012: New Evidence from a Dynamic Analysis." Accessed February 25, 2014. http://www.uiowa.edu/~stpols13/papers/Lang%20and%20Pearson%20Merkowitz%20SPPC%202013.pdf.

Lasher, Ted. n.d.(a). Notes on Conversation with Interested Party—Clinton Miller. Assemblyman Gage Papers, AB443. California State Archives, Sacramento, CA.

Lasher, Ted. n.d.(b). Notes on Conversation with Interested Party—Ellen Powell with Ajay and Miller. Assemblyman Gage Papers, AB443. California State Archives, Sacramento, CA.

Lathrop, Kyle W. 1991. "Pre-empting with Oranges: Federal Regulation of Organic Food Labeling." *Journal of Corporation Law*, 16(Summer): 885–930.

Legislative Counsel, California State Legislature. 1979. Digest of AB 443. Assembly Ways and Means Committee, AB443. California State Archives, Sacramento, CA.

Legislative Counsel, California State Legislature. 1992. Digest of AB 3246. Senate Committee on Agriculture and Water Resources, AB 645. California State Archives, Sacramento, CA.

Legislative Counsel, California State Legislature. 1995. Digest of SB 688. Assembly Committee on Agriculture and Water Resources, SB 688. California State Archives, Sacramento, CA.

Legislative Counsel, California State Legislature. 2005. Digest of SB 730. Senate Health Committee, SB 730. California State Archives, Sacramento, CA.

Legislative Counsel, California State Legislature. 2009. "Digest of AB 856 for Senate Food and Agriculture Committee." Accessed January 14, 2014. http://leginfo.legislature.ca.gov/faces/billAnalysisClient.xhtml.

Lejano, Raul, Mrill Ingram, and Helen Ingram. 2013. *The Power of Narrative in Environmental Networks*. Cambridge: MIT Press.

Light, Alfred R. 1978. "Intergovernmental Sources of Innovation in State Administration." *American Politics Quarterly*, 6(2): 147–66.

Lipson, Mark. 1990. New Mock-up of Revisions to the Bill Letter. Senate Health and Human Services Committee, AB 2012. California State Archives, Sacramento, CA.

Lipson, Mark. 1998. *Searching for the "O-Word."* Santa Cruz: Organic Farming Research Foundation.

Lipson, Mark. Interviewed by Ellen Farmer. Cultivating a Movement: Organic and Sustainable Farming, Regional History Project, University Library, University of California, Santa Cruz, June 5, 2007.

Lipson, Mark. Interview by Samantha Mosier. Tape recording/transcription, December 6, 2013.

Livingston, Gene. 2002a. Letter to Honorable Virginia Strom-Martin on August 26. Virginia Strom-Martin Papers, AB 2823. California State Archives, Sacramento, CA.

Livingston, Gene. 2002b. Letter to Members of the Senate on August 22. Assembly Committee on Agriculture, AB 2823. California State Archives, Sacramento, CA.

Lobao, Linda and Katherine Meyer. 2001. "The Great Agricultural Transition: Crisis, Change, and Social Consequences of Twentieth Century U.S. Farming." *Annual Review of Sociology*, 27: 103–24.

Lockeretz, William. 2007. "What Explains the Rise of Organic Farming?" In *Organic Farming: An International History.* Edited by William Lockeretz. Cambridge: CABI International, 1–8.

Loureiro, Maria L. and Susan Hine. 2001. "Discovering Niche Markets: A Comparison of Consumer Willingness to Pay for a Local (Colorado-Grown), Organic, and GMO-Free Product." Paper presented at American Agricultural Economics Association Meeting.

Lutz, James M. 1986. "The Spatial and Temporal Diffusion of Selected Licensing Laws in the United States." *Political Geography Quarterly*, 5(2): 141–59.

Majone, Giandomenico. 1991. Cross-National Sources of Regulatory Policymaking in Europe and the United States. *Journal of Public Policy*, 11(1): 79–106.

Manoochehri, Michael. 1998. "Certified Organic: Recent Developments in the Proposed Rule for a National 'Organic' Standard in the U.S. Food First Institute for Food and Development Policy." Accessed April 4, 2013. http://www.foodfirst.org/fr/node/213.

Martin, Orin. Interviewed by Sarah Rabkin. Cultivating a Movement: Organic and Sustainable Farming, Regional History Project. University Library, University of California, Santa Cruz, July 11, 2008.

Martin, Phillip L. and Alan L. Olmstead. 1985. "The Agricultural Mechanization Controversy." *Science*, 227(4687): 601–6.

McCarthy, Kevin. 2005. Letter to Governor Schwarzenegger on September 19. Assembly Republican Bill Analysis for Agriculture Committee. Senate Republican Committee, SB 730. California State Archives, Sacramento, CA.

McCright, Aaron M. and Riley E. Dunlap. 2011. "The Politicization of Climate Change and Polarization in the American Public's Views of Global Warming, 2001–2010." *The Sociological Quarterly*, 52: 155–94.

McEvoy, Miles. 2012. "Deputy Administrator's Corner- Organic Integrity: Past, Present, and Future." *The Organic Integrity Quarterly*, (June).

Menzel, Donald C. and Irwin Feller. 1977. "Leadership and Interaction Patterns in the Diffusion of Innovations among the American States." *The Western Political Quarterly*, 30(4): 528–36.

Merrigan, Kathleen. Interviewed by Amanda Kimble-Evans. Dig Deeper, Rodale Institute, September 14, 2012.

Michelsen, Johannes, Kennet Lynggaard, Susanne Padel, and Carolyn Foster. 2001. "Organic Farming Development and Agricultural Institutions in Europe: A Study of Six Countries." In *Organic Farming in Europe: Economics and Policy.* (Vol. 9). University of Hohenheim, Stuttgart, Germany.

Miller, Clinton. 1978. Letter to Assemblyman Gates on November 14. Assemblyman Gage Papers, AB443. California State Archives, Sacramento, CA.

Mintrom, Michael. 1997. "Policy Entrepreneurs and the Diffusion of Innovation." *American Journal of Political Science*, 41(3): 738–70.

Misra, Sukant, Chung L. Huang, and Stephen Ott. 1991. "Georgia Consumers' Preference for Organically Grown Fresh Produce." *Journal of Agribusiness*, 9(2): 53–65.

Mohr, Lawrence B. 1969. "Determinants of Innovation in Organizations." *The American Political Science Review*, 63(1): 111–26.

Mooney, Christopher Z. and Mei-Hsien Lee. 1995. "Legislative Morality in the American States: The Case of Pre-*Roe* Abortion Regulation Reform." *American Journal of Political Science*, 39(3): 599–627.

Mooney, Christopher Z. and Mei-Hsien Lee. 1999. "The Temporal Diffusion of Morality Policy: The Case of Death Penalty Legislation in the American States." *Policy Studies Journal*, 27(4): 766–80.

Mooney, Christopher Z. and Mei-Hsien Lee. 2000. "The Influence of Values on Consensus and Contentious Morality Policy: U.S. Death Penalty Reform, 1956–82." *Journal of Politics*, 62(1): 223–39.

National Agricultural Statistics Service (NASS). 2012a. "2007 Ag Atlas Maps (07-M217 and 07-M232)." United States Department of Agriculture. Accessed January 11, 2013. http://www.agcensus.usda.gov/Publications/2007/Online_Highlights/Ag_Atlas_Maps/Crops_and_Plants/.

National Agricultural Statistics Service (NASS). 2012b. "2007 Ag Atlas Maps (07-M001 and 07-M003)." United States Department of Agriculture. Accessed January 11, 2013. http://www.agcensus.usda.gov/Publications/2007/Online_Highlights/Ag_Atlas_Maps/Farms/.

National Agricultural Statistics Service (NASS). 2013. New England Cash Receipts 2012. United States Department of Agriculture New England Field Office. Accessed October 28, 2015. http://www.nass.usda.gov/Statistics_by_State/New_England_includes/Publications/cashrec08.pdf.

National Agricultural Statistics Service (NASS). 2014. Organic Farming: Results from the 2014 Organic Survey. Accessed May 15, 2016. https://www.agcensus.usda.gov/Publications/2012/Online_Resources/Highlights/Organics/2014_Organic_Survey_Highlights.pdf.

National Agricultural Statistics Service (NASS). 2015a. Northeast Maple Syrup Production. United States Department of Agriculture Northeastern Regional Field Office. Accessed October 28, 2015. http://www.nass.usda.gov/Statistics_by_State/New_England_includes/Publications/Crop_Production/NE%20Maple%20Syrup%20Production.pdf.

National Agricultural Statistics Service (NASS). 2015b. New England 2014 Organic Survey Results. United States Department of Agriculture New England Field Office. Accessed October 28, 2015. http://www.nass.usda.gov/Statistics_by_State/New_England_includes/Publications/Current_News_Release/NG2014OrganicSurveyResults.pdf.

National Center for Health Statistics. 2012. "Centers for Disease Control and Prevention, United States Department of Health and Human Services. Health, United States, 2011: With Special Feature in Socioeconomic Status and Health." Washington, D.C: United States Department of Health and Human Services.

National Organic Program (NOP). 2009. "NOP 5012 Approval of Liquid Fertilizers Internal Rev01 07 22 11." Accessed February 25, 2014. Agricultural Marketing Service, United States Department of Agriculture. http://www.ams.usda.gov/AMSv1.0/getfile?dDocName=STELPRDC5087119.

National Organic Program (NOP). 2011. "Information for State Organic Programs." Agricultural Marketing Service, United State Department of Agriculture. Accessed February 25, 2014. http://www.ams.usda.gov/AMSv1.0/ams.fetchTemplateData.do?template=TemplateN&navID=NationalOrganicProgram&leftNav=NationalOrganicProgram&page=NOPStateOrganicPrograms&description=State%20Organic%20Programs%20(SOPs)&acct=nopgeninfo.

National Organic Program (NOP). 2012. "National Organic Standards Board." Agricultural Marketing Service, United States Department of Agriculture. Accessed April 5, 2013. http://www.ams.usda.gov/AMSv1.0/NOSB.

National Organic Program (NOP). 2013. "Organic Integrity Database." Accessed December 1, 2013. https://organic.ams.usda.gov/integrity/Certifiers/CertifiersLocationsSearchPage.aspx.

National Organic Program (NOP). 2013a. "Information for International Trade Partners." Agricultural Marketing Service, United States Department of Agriculture. Accessed April 5, 2013. http://www.ams.usda.gov/AMSv1.0/ams.fetchTemplateData.do?template=TemplateJ&leftNav=NationalOrganicProgram&page=NOPInternationalAgreements&description=International%20Agreements&acct=nopgeninfo.

National Organic Program (NOP). 2013b. "2012 List of Certified USDA Organic Operations." Agricultural Marketing Service, United States Department of Agriculture. Accessed April 5, 2013. http://apps.ams.usda.gov/nop/.

National Organic Program (NOP). 2013c. "USDA Accredited Certifying Agents (ACAs)." Agricultural Marketing Service, United States Department of Agriculture. Accessed April 5, 2013. http://www.ams.usda.gov/AMSv1.0/ams.fetchTemplateData.do?template=TemplateJ&navID=NationalOrganicProgram&leftNav=NationalOrganicProgram&page=NOPACAs&description=USDA%20Accredited%20Certifying%20Agents&acct=nopgeninfo.

National Organic Standards Board (NOSB). n.d. "Former Members." Accessed November 1, 2015. http://www.ams.usda.gov/rules-regulations/organic/nosb/former-members.

National Park Service (NPS). n.d. Agriculture and Industry. Accessed October 27, 2015. At http://www.nps.gov/nr/travel/centralvermont/agind1.htm.

Negin, Elliot. 1996. "The Alar 'Scare' Was for Real: And So is That 'Veggie Hate-Crime' Movement." *Columbia Journalism Review*. Accessed February 25, 2014. http://www.pbs.org/tradesecrets/docs/alarscarenegin.html.

Nicholson-Crotty, Sean. 2004. "The Politics and Administration of Privatization: Contracting Out for Corrections Management in the United States." *Policy Studies Journal*, 32: 41–57.

Niggli, Urs. 2007. "FiBL and Organic Research in Switzerland." In *Organic Farming: An International History*. Edited by William Lockeretz. Cambridge: CABI International.

NOFA-VT. 2015. Certified Organic Agriculture in Vermont. Obtained via Personal Communications October 2015.

NOFA-VT Staff Member. Interviewed by Samantha Mosier. Tape recording/transcription. October 28, 2015.

Norberg-Hodge, Helena, Peter Goering, and John Page. 1993. *From the Ground Up: Rethinking Industrial Agriculture*. London: Zed Books.

Obach, Brian K. 2015. *Organic Struggle: The Movement for Sustainable Agriculture in the United States*. Cambridge, MA: MIT Press.

Office of Assemblywoman Jackie Speier. 2005. SB 730 Background Information. Assembly Agriculture Committee, SB 730. California State Archives, Sacramento, CA.

Office of Environmental Health Hazard Assessment. 2013. Chemicals Known to the State to Cause Cancer or Reproductive Toxicity July 26, 2013. Environmental Protection Agency, State of California.

Office of Governor Peter Shumlin. 2011. Dairy Still Growing in Vermont. Press Release.

Olmstead, Alan L. and Paul W. Rhode. 2003. "The Evolution of California Agriculture 1850–2000." In *California Agriculture: Dimensions and Issues*. Edited by Jerry Siebert. Giannini Foundation of Agricultural Economics, University of California.

Onozaka, Yuko, Gretchen Nurse, and D. Thilmany-McFadden. 2011. "Defining Sustainable Food Market Segments: Do Motivations and Values Vary by Shopping Locale?" *American Journal of Agricultural Economics*, 93(2): 583–89.

Onozaka, Yuko and Dawn Thilmany-McFadden. 2011. "Does Local Labeling Complement or Compete with Other Sustainable Labels? A Conjoint Analysis of Direct and Joint Values for Fresh Produce Claims." *American Journal of Agricultural Economics*, 93(3): 689–702.

Oregon Tilth. N.A. "History." Accessed April 2, 2013. http://tilth.org/about/history.

Organic Trade Association (OTA). 2011. *OTA's 2011 Organic Industry Survey*. Accessed February 25, 2014. http://www.ota.com/pics/documents/2011OrganicIndustrySurvey.pdf.

Organic Trade Association (OTA). 2016. *OTA's 2016 Organic Industry Survey*. Accessed December 15, 2016. https://www.ota.com/resources/organic-industry-survey.

Parsons, Bob. 2015. "Vermont's Dairy Sector: Is There a Sustainable Future for the 800 lb. Gorilla?" *Opportunities for Agriculture Working Paper Series (Food System Research Collaboration of University of Vermont Rural Studies Center)*: 1(4). Accessed October 28, 2015. http://www.uvm.edu/crs/reports/working_papers/WorkingPaperParsons-web.pdf.

Peck, W. Taylor. 2002. Letter to Assembly Member Strom-Martin from Nub Circus on August 16. Virginia Strom-Martin Papers, AB 2823. California State Archives, Sacramento, CA.

Place-Based Landscape Analysis and Community Engagement (PLACE). 2015. "Cultural Landscape: Rural Landscape." Accessed October 27, 2015. http://www.uvm.edu/place/analyze/rural.php.

Popoff, Mischa. 2010. *Is It Organic? The Inside Story of Who Destroyed the Organic Industry, Turned It Into a Socialist Movement and Made Million$ in the Process.* United States: Xulon Press.

Price, Charlene C. 1985. "Eating 'Natural' Gains Popularity." *National Food Review*, 28:14–18.

Ragland, Edward, and Debra Tropp. 2009. USDA National Farmers Market Manager Survey. US Department of Agriculture, Agricultural Marketing Service. Accessed January 31, 2016. http://dx.doi.org/10.9752/MS037.05-2009.

Rawson, Jean M. Congressional Research Service. 1998. *Organic Foods and the Proposed Federal Certification and Labeling Program.* Accessed April 4, 2013. http://crs.ncseonline.org/NLE/CRSreports/Agriculture/ag-54.cfm#4.

Rawson, Julie, Elizabeth Henderson, Grace Gershuny, Robert Cox, and Bill Duesing. 2011. NOFA History Conference Presentation. Accessed October 15, 2016. http://www.slideshare.net/bduesing/nofa-history-2011-conference-presentation.

Raynolds, Laura T. 2000. "Re-embedding Global Agriculture: The International Organic and Fair Trade Movements." *Agriculture and Human Values*, 17(3): 297–309.

Regens, James L. 1980. "State Policy Responses to the Energy Issue: An Analysis of Innovation." *Social Science Quarterly*, 61(1): 44–57.

Renner, Lacy Boggs. 2012. "Major Natural Food Brand Opposing GMO Labeling." Accessed April 4, 2013. http://www.organicauthority.com/foodie-buzz/natural-foods-fighting-gmo-labeling.html.

Rich, Deborah. 2008. "Organic and Sustainable Up for Review . . . Again." *OFRF Information Bulletin*, 16: 6–13.

Rigby, Dan and Sophie Brown. 2003. "Organic Food and Global Trade: Is the Market Delivering Agricultural Sustainability?" The University of Manchester School of Economic Studies Discussion Paper Series, Number 0326.

Roberts, Phillip. 2013. "Boll Weevil." *Georgia Encyclopedia.* Accessed February 25, 2014. http://www.georgiaencyclopedia.org/articles/business-economy/boll-weevil.

Rodale Institute. n.d. "Mission and History." Accessed April 2, 2013. http://rodaleinstitute.org/about-us/mission-and-history/.

Rogers, Debbie. Senate Republic Office of Policy. 2002. Whip Comments. Senator Representative Polanco Files, AB 2823. California State Archives, Sacramento, CA.

Rogers, Everett M. 1957. "A Conceptual Variable Analysis of Technological Change." PhD diss. Iowa State College.

Rogers, Everett M. 2003. *The Diffusion of Innovations*. 5th ed. New York: Free Press.

Rose, Douglas D. 1973. "National and Local Forces in State Politics: The Implications of Multi-Level Policy Analysis." *The American Political Science Review*, 67: 1162–73.

Rosenbaum, Nelson M. 1976. *Land Use and the Legislature: The Politics of State Innovation*. Washington, D.C.: The Urban Institute.

Rural Vermont. 2015. "History of Rural Vermont." Accessed October 28, 2015. http://www.ruralvermont.org/about-us/history/.

Savage, Robert L. 1985. "Diffusion Research Traditions and the Spread of Policy Innovations in a Federal System." *Publius*, 15(4): 1–27.

Schlosser, Eric. 2001. *Fast Food Nation: The Dark Side of the All-American Meal*. New York: Houghton Mifflin.

Schwieder, Dorothy. 1990. "Agricultural Issues in the Middle West, 1865–1900." *Agriculture and National Development*. Edited by Lou Ferleger. Ames, Iowa: Iowa State University Press.

Scrowcroft, Bob. Interviewed by Irene Reti. Cultivating a Movement: Organic and Sustainable Farming. Regional History Project. University Library, University of California, Santa Cruz, December 18, 2007.

Sharkansky, Ira. 1969. "The Utility of Elazar's Political Culture: A Research Note." *Polity*, 2(1): 68–83.

Sharman, J. S. 2010. Dysfunctional Policy Transfer in National Tax Blacklists. *Governance*, 23(4): 623–39.

Shipan, Charles R. and William R Lowry. 2001. "Environmental Policy and Party Divergence in Congress." *Political Research Quarterly*, 54(2): 245–63.

Shipan, Charles R. and Craig Volden. 2006. "Bottom-Up Federalism: The Diffusion of Antismoking Policies from U.S. Cities to States." *American Journal of Political Science*, 52(4): 840–57.

Shipan, Charles R. and Craig Volden. 2008. "The Mechanisms of Policy Diffusion." *American Journal of Political Science*, 52(4): 840–57.

Shipan, Charles R. and Craig Volden. 2012. "Policy Diffusion: Seven Lessons for Scholars and Practitioners." *Public Administration Review*, 7(6): 788–96.

Sinclair, Upton. 1906. *The Jungle*. United States: Doubleday, Jabber & Company.

Sligh, Michael and Thomas Cierpka. 2007. "Organic Values." In *Organic Farming: An International History*. Edited by William Lockeretz. Cambridge: CABI International.

Soule, Sarah A. 1999. "The Diffusion on an Unsuccessful Innovation." *Annals of the American Academy of Political and Social Science*, 566: 120–31.

Squire, Peverill. 2007. "Measuring State Legislative Professionalism: The Squire Index Revisited." *State Politics and Policy Quarterly*, 7(2): 211–27.

Stern, Joseph. 2002. Letter to Assembly Member Strom-Martin from Organic Ingredients Inc. on August 16. Virginia Strom-Martin Papers, AB 2823. California State Archives, Sacramento, CA.

Stevens, Will. 1990. Testimony on S.277 for the Vermont Senate Agriculture Committee. State of Vermont Archives A116–00381.

Stinner, Deborah H. 2007. "The Science of Organic Farming." In *Organic Farming: An International History*. Edited by William Lockeretz. Cambridge: CABI International.

Strom-Martin, Virginia. 2002a. Letter to the Honorable Gray Davis. Virginia Strom-Martin Papers, AB 2823. California State Archives, Sacramento, CA.

Strom-Martin, Virginia. 2002b. AB 2823-Organic Products (Strom-Martin) File Item 153. Virginia Strom-Martin Papers, AB 2823. California State Archives, Sacramento, CA.

Sumner, Daniel, Jose E. Bervejillo, and Nicolai V. Kuminoff. 2003. "The Measure of California Agriculture and its Importance in the State's Economy." In *California Agriculture: Dimensions and Issues*. Edited by Jerry Siebert. Giannini Foundation of Agricultural Economics, University of California.

Sustainable Food Advocate. Interviewed by Samantha Mosier. Tape recording/transcription. October 30, 2013.

Tarde, Gabriel. 1903. *The Laws of Imitation*. New York: University of Chicago Press.

Taylor, Kathryn C. 2013. "Peaches." *Georgia Encyclopedia*. Accessed January 11, 2014. http://www.georgiaencyclopedia.org/articles/business-economy/peaches.

Timmons, Gay, COPA 2003 Task Force. 2002a. Letter to Assembly Agriculture Committee on April 11. Assembly Committee on Agriculture, AB 2823. California State Archives, Sacramento, CA.

Timmons, Gay, COPA 2003 Task Force. 2002b. Letter to Assembly Member Virginia Strom-Martin on August 16. Virginia Strom-Martin Papers, AB 2823. California State Archives, Sacramento, CA.

Tindall, George B. 1972. *Disruption of Solid South*. Athens: University of Georgia Press.

U.S. Census Bureau. 1913. *Thirteenth Census of the United States Taken In the Year 1910 Vol. VII: Agriculture 1909 and 1910.* Department of Commerce.

U.S. Census Bureau. 2013. "American Community Survey, 5-Year Estimate by County." Accessed January 15, 2014. http://factfinder2.census.gov/.

U.S. Census Bureau. n.d. "Historical Income Tables: Household, Table H-6 (Regions by Median and Mean Income)." Accessed February 25, 2014 http://www.census.gov/hhes/www/income/data/historical/household/.

United States Department of Agriculture (USDA) Study Team (on Organic Farming). 1980. "Report and Recommendations on Organic Farming." Accessed April 4, 2013. http://www.nal.usda.gov/afsic/pubs/USDAOrgFarmRpt.pdf.

United States Department of Agriculture (USDA) Ag Census. 2012. "Vermont State Data on Land: 2012 and 2007." Accessed October 27, 2015. http://www.agcensus.usda.gov/Publications/2012/Full_Report/Volume_1,_Chapter_1_State_Level/Vermont/st50_1_008_008.pdf.

University of California-Berkeley Library (UCB-Library). n.d. "California History of Agriculture." Accessed February 25, 2014. http://harvest.mannlib.cornell.edu/node/11.

University of California Cooperative Extension, UC Small Farm Program (UCSFP). 2014. "About Us." Accessed February 25, 2014. http://sfp.ucdavis.edu/overview/about/.

University of Georgia Cooperative Extension. 2012. "Ag Stats." Accessed February 25, 2014. http://extension.uga.edu/agriculture/ag-stats-agencies/ag-stats/.

University of Georgia Libraries. n.d. "Georgia History of Agriculture." Accessed January 11, 2014. http://harvest.mannlib.cornell.edu/node/15.

Vermont Secretary of State. n.d. "Vermont Election Archive." Accessed December 1, 2016. http://vtelectionarchive.sec.state.vt.us.

Vermont Senate Agriculture Committee. 1990. Committee Minutes 4/9/90. State of Vermont Archives.

Vermont Senate Agriculture Committee. 1998a. Committee Minutes 4/10/98. State of Vermont Archives A116–00253.

Vermont Senate Agriculture Committee. 1998b. Committee Minutes Recording 4/10/98. State of Vermont Archives Cassette T-2162 4/10.

Vermont Senate Agriculture Committee. 2003. Committee Minutes 1/17/2003. State of Vermont Archives A116–00381.

Vermont Senator. Interviewed by Samantha Mosier. Tape recording/transcription. September 30, 2015.

Vogt, Gunter. 2007. "The Origins of Organic Farming." In *Organic Farming: An International History*. Edited by William Lockeretz. Cambridge: CABI International.

Voss, Henry J. and Richard W. Nutter. 1989. Letter from CDFA to the Honorable Delaine Eastin. Assembly Committee on Government Efficiency and Consumer Protection, AB 2012. California State Archives, Sacramento, CA.

Walker, Jack L. 1969. "The Diffusion of Innovations among the American States." *The American Political Science Review*, 63(3): 880–99.

Walker, Jack L. 1973. "Comment: Problems in Research on the Diffusion of Policy Innovations." *The American Political Science Review*, 67(4): 1186–91.

Weber, Reagan. 2010. "Senate Passes GA Taxpayer Relief Act." Accessed February 25, 2014. http://senatepress.net/senate-passes-ga-taxpayer-relief-act.html.

Weissert, Carol S. and Daniel Scheller. 2008. "Learning from the States? Federalism and National Health Policy." *Public Administration Review*, 68: S162–74.

Welch, Susan and Kay Thompson. 1980. "The Impact of Federal Incentives on State Policy Innovation." *American Journal of Political Science*, 24(4): 715–29.

Weyland, Kurt G. 2004. *Learning from Foreign Models in Latin American Policy Reform.* Washington DC: Woodrow Wilson Center Press.

Weyland, Kurt G. 2007. *Bounded Rationality and Policy Diffusion*. Princeton, NJ: Princeton University Press.

Wilkie, Barbara. 2002. Letter to Assembly Member Strom-Martin from Environmental Health Network on August 19. Virginia Strom-Martin Papers, AB 2823. California State Archives, Sacramento, CA.

Winders, Bill and James C. Scott. 2009. *The Politics of Food Supply: U.S. Agricultural Policy in the World Economy*. New Haven: Yale University Press.

Wonnacott, Enid. 1989. September 11. Testimony to U.S. Senate Agriculture Committee Hearing in Montpelier, Vermont. Personal Communications.

Wonnacott, Enid. Personal Communications by Samantha Mosier. E-mail and Historical Documentation. November 2 and 3, 2015.

Wonnacott, Enid. n.d. Consumer as Co-Producers: Influencing the Tipping Point towards Thriving Local Food Systems. Personal Communications.

Woolf, Art. 2015. "A Look at What Jobs are Growing in VT." *Burlington Free Press.* Accessed October 26, 2015. http://www.burlingtonfreepress.com/story/money/2015/08/26/look-jobs-growing-vt/32433835/.

Youngberg, Garth, Neill Schaller, and Kathleen Merrigan. 1993. "The Sustainable Agriculture Policy Agenda in the United States: Politics and Prospects." In *Food for the Future: Conditions and Contradictions*. Edited by Patricia Allen. New York: Wiley, 295–318.

Yue, Chengyan and Cindy Tong. 2009. "Organic or Local? Investigating Consumer Preference for Fresh Produce Using a Choice Experiment with Real Economic Incentives." *HortScience*, 44(2): 366–71.

Zepeda, Lydia and David Deal. 2009. "Organic and Local Food Consumer Behavior: Alphabet Theory." *International Journal of Consumer Studies*, 33(6): 697–705.

Index

Page references for figures and tables are italicized.

About the Author

Samantha L. Mosier is an assistant professor at Missouri State University with expertise in subnational environmental policy and public policy theory and analysis. Her research, which utilizes both quantitative and qualitative evaluation, focuses on the dynamics underlying the adoption and change of innovative policies with specific focus on local sustainability initiatives and sustainable agriculture.

www.ingramcontent.com/pod-product-compliance
Lightning Source LLC
LaVergne TN
LVHW101447200926
848483LV00001B/73